T0183068

SpringerBriefs in Applied Sciences and Technology

Manufacturing and Surface Engineering

Series editor

Joao Paulo Davim, Aveiro, Portugal

More information about this series at http://www.springer.com/series/10623

Kapil Gupta · Neelesh K. Jain
R.F. Laubscher

Hybrid Machining Processes

Perspectives on Machining and Finishing

 Springer

Kapil Gupta
School of Mechanical and Industrial
  Engineering
University of Johannesburg
Johannesburg
South Africa

R.F. Laubscher
School of Mechanical and Industrial
  Engineering
University of Johannesburg
Johannesburg
South Africa

Neelesh K. Jain
Discipline of Mechanical Engineering
Indian Institute of Technology Indore
Indore
India

ISSN 2191-530X   ISSN 2191-5318   (electronic)
SpringerBriefs in Applied Sciences and Technology
ISSN 2365-8223   ISSN 2365-8231   (electronic)
Manufacturing and Surface Engineering
ISBN 978-3-319-25920-8   ISBN 978-3-319-25922-2   (eBook)
DOI 10.1007/978-3-319-25922-2

Library of Congress Control Number: 2015953270

Springer Cham Heidelberg New York Dordrecht London
© The Author(s) 2016
This work is subject to copyright. All rights are reserved by the Publisher, whether the whole or part of the material is concerned, specifically the rights of translation, reprinting, reuse of illustrations, recitation, broadcasting, reproduction on microfilms or in any other physical way, and transmission or information storage and retrieval, electronic adaptation, computer software, or by similar or dissimilar methodology now known or hereafter developed.
The use of general descriptive names, registered names, trademarks, service marks, etc. in this publication does not imply, even in the absence of a specific statement, that such names are exempt from the relevant protective laws and regulations and therefore free for general use.
The publisher, the authors and the editors are safe to assume that the advice and information in this book are believed to be true and accurate at the date of publication. Neither the publisher nor the authors or the editors give a warranty, express or implied, with respect to the material contained herein or for any errors or omissions that may have been made.

Printed on acid-free paper

Springer International Publishing AG Switzerland is part of Springer Science+Business Media
(www.springer.com)

Preface

The more stringent requirements for enhanced quality of engineered products, especially those of miniature size and made from difficult-to-machine materials, have largely been responsible for the development and subsequent wide use of hybrid machining processes (HMPs). HMPs are variants of advanced machining processes (AMPs) that are combinations of either one or more specific AMP or an AMP combined with a conventional metal working process (usually some form of machining process) to achieve results that would not be possible with the individual constituent processes in isolation.

HMPs may also include assistance of an additional external energy source or fluid and/or abrasive media to enhance material removal rate and/or surface quality. HMPs are slowly but surely finding their way into mainstream manufacturing globally due to the associated high-quality machining and finishing capabilities. Micro-machining and micro-finishing are the two most important application domains where HMPs have demonstrated significant advantages. Their ability to machine extremely hard, brittle and difficult-to-cut materials and to produce improved surface integrity aspects are the key advantages.

The main objective of this book is to provide a wider perspective on some of the most important HMPs used for micro-machining and finishing purposes and to provide a resource for better understanding of the underlying working principles of the different processes. This book aims to address the needs of researchers, students and professionals in the fields of mechanical and manufacturing engineering as a first step into the world of hybrid machining processes.

Chapter 1 of this book presents an introduction, a brief overview and a detailed classification scheme of HMPs along with reference to some important applications. Electrochemical hybrid machining processes are presented in Chap. 2. This is followed in Chap. 3 with an introduction to thermal hybrid machining processes. Chapter 4 introduces various assisted type HMPs including vibration-assisted, heat-assisted, abrasive-assisted and magnetic field-assisted hybrid machining processes. The content of the various chapters is arranged to include a basic introduction of the process in question, equipment details, working principles, significant process parameters and important applications. References are made throughout to the current and previous research work conducted as related to the individual processes.

Kapil Gupta
Neelesh K. Jain
R.F. Laubscher

Contents

Chapter 1
Overview of Hybrid Machining Processes

Abstract Most of the unconventional or advanced machining processes (AMPs) were developed after the World War II in response to the challenges of machining and/or finishing unusual shapes and/or sizes in the advanced and difficult-to-machine (DTM) materials or for those applications where use of conventional machining processes is technically unfeasible, detrimental to the useful material properties, uneconomical and unproductive. But, further development of advanced materials and emphasis on miniaturization and ultra-precision machining and/or finishing requirements exposed limitations of individual AMPs. These challenges are being met with the development of hybrid machining processes (HMPs) by either combining two AMPs or one AMP with a conventional machining process to exploit their capabilities in a single process and simultaneously overcome their individual limitations. This chapter introduces AMPs, defines and classifies HMPs and refers to their applications.

Keywords Advanced machining process (AMP) · Hybrid machining process (HMP) · Combined HMP · Assisted HMP

1.1 Evolution of Advanced Machining Processes

Stiff market competition, increasing demand for precise and high-quality products and their improved performance in all industrial segments have led to the development of an ever-growing variety and quality of advanced materials which are difficult-to-machine (DTM) such as super alloys, some non-ferrous alloys (i.e. titanium-based alloys), ceramics, composites, polymers, functionally graded materials, shaper memory alloys and nano-materials. These materials may possess exceptional properties such as high strength and stiffness at elevated temperatures, extreme hardness, high brittleness, high strength-to-weight ratio, high corrosion and oxidation resistance, and chemical inertness, which usually results in superior product performance but may also make their machining and/or finishing

© The Author(s) 2016
K. Gupta et al., *Hybrid Machining Processes*,
Manufacturing and Surface Engineering,
DOI 10.1007/978-3-319-25922-2_1

challenging. These advanced materials play an important role in modern manu-facturing industries especially in automobiles, aviation, cutting tools, dies, mechatronics, nuclear, biomedical, computers, etc. Using conventional machining processes for such materials may result in excessive machining cost, and degra-dation of some of their useful properties [1, 2]. Therefore, the manufacturing industry is facing challenges mainly from the (i) continuous development and use of DTM materials in various key and strategic industrial applications; (ii) fabrication of MEMS and NEMS; (iii) precise machining of complicated shapes and/or sizes; (iv) machining at micro- and nano-level; (v) machining of inaccessible areas; (vi) various hole drilling requirements such as non-circular holes, deeps holes, micro-holes and large number of holes in close proximity; and (vii) machining of low rigidity structures. Figure 1.1 depicts some of those machining challenges.

AMPs have evolved in response to meet these challenges. AMPs are referred to as 'advanced' manufacturing processes because they utilize new forms of energy

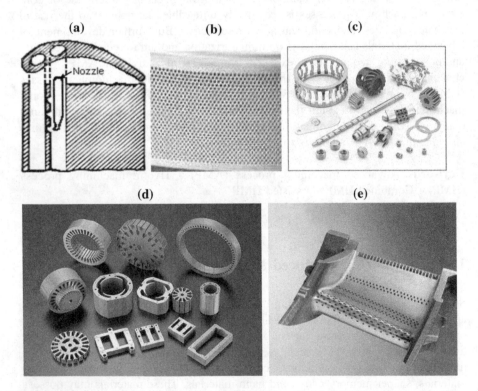

Fig. 1.1 Some challenges which can be met by AMPs only (**a**) drilling an inaccessible hole (normal to the wall in the present case); (**b**) drilling a large number closely spaced holes; (**c**) machining at nano-, micro- and meso-level (*courtesy* Techniks Inc.); (**d**) machining of parts with typical intricate features (*courtesy* AGS-TECH Inc.); and (**e**) machining deep holes on a curved surface, i.e. turbine blade (*courtesy* Industrial Laser LLC)

Table 1.1 Classification of AMPs according to the energy source [1, 2]

Type of energy	AMP	Energy source	Tool	Transfer media	Mechanism of material removal
Mechanical	USM	Ultrasonic vibration	Sonotrode	Abrasive slurry	Erosion or abrasion
	AJM	Pneumatic pressure	Abrasive jet	Air	
	WJM	Hydraulic pressure	Water jet	Air	
	AWJM	Hydraulic pressure	Abrasive water jet	Air	
	IJM	Hydraulic pressure	Ice jet	Air	
	AFM	Hydraulic pressure	Abrasives	Putty	
Chemical	CHM	Corrosive agent	Mask	Etchant	Chemical dissolution
Electrochemical	ECM	High current	Electrode	Electrolyte	Anode dissolution through ion displacement
Thermal	EDM	High voltage	Electrode	Dielectric	Melting and vaporization
	EBM	Ionized material	Electron beam	Vacuum	
	IBM	Ionized material	Ion beam	Atmosphere	
	LBM	Amplified light	Laser beam	Air	
	PAM	Ionized material	Plasma jet	Plasma	

AFM abrasive flow machining; *AJM* abrasive jet machining; *AWJM* abrasive water jet machining; *CHM* chemical machining; *EBM* electron beam machining; *ECM* electrochemical machining; *EDM* electric discharge machining; *IBM* ion beam machining; *IJM* ice jet machining; *LBM* laser beam machining; *PAM* plasma arc machining; *USM* ultrasonic machining; *WJM* water jet machining

and tools. Table 1.1 presents a classification of AMPs according to the type of energy source giving details of mechanism of material removal, energy source, cutting tool, and transfer media used in each process.

1.2   Need and Introduction of Hybrid Machining Processes

The AMP categories referred to in Table 1.1 each presents its unique and inherent limitations; for example, (i) all the mechanical AMPs using abrasives (i.e. USM, AJM, AWJM, AFM) require hardness of abrasive particles more than the workpiece material at the actual temperature of machining. Moreover, these AMPs also impart mechanical damage (i.e. residual stresses, micro-cracks and hardness alternations) to the workpiece surface; (ii) ECM can only be used for electrically conducting materials, while CHM can only be used for chemically reactive materials. Both these AMPs may induce chemical damage (i.e. inter-granular attack/oxidation/corrosion, preferential dissolution of micro-constituents, contamination, embrittlement and corrosion) to the workpiece surface; and (iii) thermal AMP may impart thermal damage (i.e. heat affected zone, recrystallization, recast layer and splattered particles). Moreover, costs for equipment, operation and maintenance of thermal AMPs are high. However, hybridizing an AMP with another AMP or with a conventional machining/finishing process to develop an HMP enables it to overcome these limitations.

Probably, the main objective of an HMP is to use the combined or mutually enhanced advantages to obtain a '1 + 1 = 3' effect, i.e. the positive effect of an HMP is larger than the simple sum of the individual advantages of the constituent processes [3, 4] and may also simultaneously eliminate or minimize the adverse effects that the constituent processes may have when used individually. Additionally, HMPs are able (i) to meet some of the ultra-precision machining requirements; (ii) to meet high productivity requirements for the components made of advanced, exotic and difficult-to-machine materials; (iii) to meet the challenges of stringent shape and size requirements; and (iv) to achieve enhanced surface quality and tolerance requirements with good productivity.

In an HMP, the processes/energy sources should interact more or less in the same machining zone and at the same time. This intensifies the machining operation and results in low tool wear, better work surface integrity, increased productivity, better product quality and influentially increases the overall efficiency. For example, both ECM and EDM can be used only for electrically conducting materials and have certain limitations on productivity and product quality, but their combination in a HMP known as electrochemical discharge/arc machining (ECDM or ECAM) or electroerosion dissolution machining (EEDM) may have 5 to 50 times greater productivity than that of ECM or EDM, and it can be used for both electrically conducting and non-conducting materials [1, 2].

1.3   Classification of Hybrid Machining Processes

Figure 1.2 shows a detailed taxonomy of HMPs. The HMPs can be broadly classified into the following two categories [3–6]:

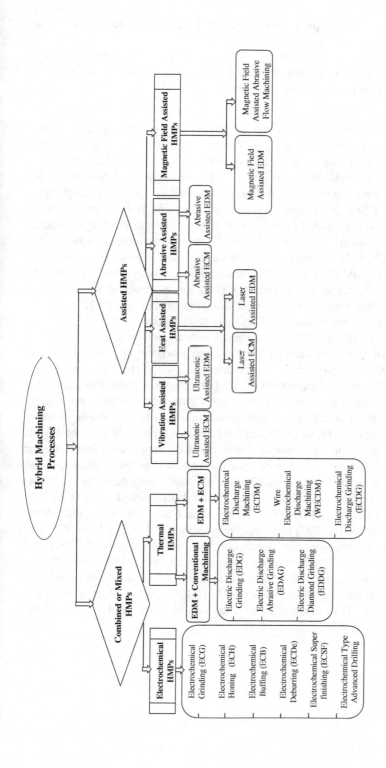

Fig. 1.2 A taxonomy of some important hybrid machining processes

1. **Combined or mixed-type** processes in which all constituent processes are directly involved in the material removal;
2. **Assisted-type** processes in which only one of the participating processes directly remove material, while the other only assists in removal by having a positive effect on the conditions of machining.

Combined or mixed HMPs are those processes in which two or more energy sources/tools/mechanisms are combined and have a synergetic effect on the material removal process. They can further be categorized as electrochemical HMPs and thermal HMPs. Electrochemical grinding (ECG), electrochemical honing (ECH), electrochemical deburring (ECDe), electrochemical buffing (ECB), electrochemical superfinishing (ECSF) and electrochemical-type advanced drilling processes are some of the important electrochemical-type HMPs. Electric discharge grinding (EDG), electric discharge abrasive grinding (EDAG) and electric discharge diamond grinding (EDDG) are some important thermal HMPs involving a combination of a thermal AMP (i.e. EDM) and a conventional finishing process. Electrochemical discharge machining (ECDM) or electroerosion dissolution machining (EEDM), wire electrochemical discharge machining (WECDM) and electrochemical discharge grinding (ECDG) are some important thermal HMPs in which a thermal AMP (i.e. EDM) is combined with an electrochemical AMP.

In the *assisted HMPs*, the material is removed by a primary process which is an AMP, while a secondary process provides assistance by utilizing either high-frequency vibrations, heat, abrasives, fluid or a magnetic field. This assistance works to intensify the material removal mechanism, removing burrs and recast layers from the work surface, ejecting the molten material from the machining zone or finishing the workpiece. These combined- and assisted-type HMPs can be further classified based on the energy source and/or mechanism of material removal, and/or the source of assistance such as vibration-assisted, heat-assisted and medium-assisted HMPs. Ultrasonic-assisted ECM (USECM) and ultrasonic-assisted EDM (USEDM) are the two main types of vibration-assisted HMPs; laser-assisted ECM (LAECM) and laser-assisted EDM (LAEDM) are the two main types of heat-assisted HMPs; abrasive-assisted ECM (AECM) and abrasive-assisted EDM (AEDM) are the main abrasive-assisted HMPs, while in conclusion, magnetic-field-assisted EDM and magnetic-field-assisted AFM are the two important magnetic-field-assisted HMPs.

There are still some other HMPs based on the assistance from rheological fluids such as magnetorheological fluid-assisted AFM (MRFA-AFM), electrorheological fluid-assisted USM (ERFA-USM) and electrorheological fluid-assisted EDM (ERFA-EDM).

1.4 Applications of Hybrid Machining Processes

Although the initial developments of the individual HMPs were material, shape or application specific, HMPs may in general be used for wide variety of applications. Most HMPs are capable of providing good surface quality with average surface roughness value up to 0.5 µm, good surface integrity, defect-free and stress-free surfaces with improved productivity. HMPs are also used for ultra-precision machining (i.e. surface roughness up to 1 nm and tolerances up to 10 nm); machining and/or finishing of nano-/micro-sized components such as 3D micro-features, micro-holes and cavities, and machining and/or finishing of advanced and difficult-to-machine materials. The potential applications, capabilities, working principle and mechanisms of the different types of HMPs are discussed as appropriate in subsequent chapters of this book.

References

1. Pandey PC, Shan HS (1980) Modern machining processes. McGraw Hill Education (India) Pvt Ltd., New Delhi
2. Jain VK (2002) Advanced machining processes. Allied Publishers Pvt Ltd., New Delhi
3. Lauwers B, Klocke F, Klink A, Tekkaya AE, Neugebauer R, Mcintosh D (2014) Hybrid processes in manufacturing. Ann CIRP 63:561–583
4. Lauwers B (2011) Surface integrity in hybrid machining processes. Procedia Eng 19:241–251
5. Kozak J, Rajurkar KP (2000) Hybrid machining process evaluation and development. In: Proceedings of 2nd international conference on machining and measurements of sculptured surfaces, Keynote Paper, Krakow, pp 501–536
6. Zhu Z, Dhokia VG, Nassehi A, Newman ST (2013) A review of hybrid manufacturing processes—state of the art and future perspectives. Int J Comput Integr Manuf 26(7):596–615

Chapter 2
Electrochemical Hybrid Machining Processes

Abstract This chapter introduces electrochemical HMPs developed by combining electrochemical machining (ECM) with conventional finishing processes such as grinding, mechanical honing, buffing, deburring and superfinishing with an objective to overcome their limitations and to enhance their finishing capabilities. Electrochemical HMPs are generally recognized as micro-finishing processes which are used to finish gears, cylinders, micro-shafts, and other micro-, meso- and macro-engineered components in order to improve surface finish and/or to correct micro-irregularities. The current chapter also presents brief details of electrochemical-type advanced drilling processes which have been developed to meet specialized drilling requirements.

Keywords ECM · Electrolyte · Dissolution · Finishing · Burr

Electrochemical machining (ECM) is a 'copying' process in which an approximate complementary image of a cathode tool is reproduced on the anode workpiece through controlled electrolytic dissolution (ED) by circulating an appropriate electrolyte in the inter-electrode gap (IEG). ECM is a self-regulating process due to the ED being governed by Faraday's laws of electrolysis which mandates that the material removal rate (MRR) is inversely proportional to the inter-electrode gap and directly proportional to electrolyte conductivity and DC voltage applied across the inter-electrode gap.

Being a non-contact machining process, ECM offers many advantages, namely (i) process performance being independent of mechanical properties of the workpiece material which basically implies that materials of any hardness and/or toughness can be machined; (ii) production of largely stress-free surface; (iii) good surface finish and integrity; (iv) higher productivity due to higher MRRs. It does, however, also suffer from inherent significant limitations, namely (i) passivation of workpiece surface by non-conducting metal oxide layer formation due to the evolution of oxygen gas at the anode; (ii) applicability to only electrically conducting materials; (iii) corrosion of machining elements and surroundings due to the use of an electrolyte; (iv) chemical damage caused to workpiece surface; (v) the dependence of accuracy on the inter-electrode gap which requires efficient flushing

© The Author(s) 2016   9
K. Gupta et al., *Hybrid Machining Processes*,
Manufacturing and Surface Engineering,
DOI 10.1007/978-3-319-25922-2_2

of sludge; and (vi) process performance being sensitive to electrolyte characteristics. ECM is invariably used in the machining of difficult-to-machine but electrically conducting materials, which are extensively used in automobiles, aerospace, defence, cutting tools, dies and moulds and biomedical applications. However, conventional finishing processes such as grinding, mechanical honing, buffing and superfinishing are inexpensive and give good surface quality, but the use of abrasives imparts some inherent limitations such as (i) high tool wear; (ii) low productivity; and (iii) mechanical damage to the finished surface. One of the recent developments in the field of ECM is to combine it with some conventional machining/finishing or AMP to develop an electrochemical HMP (ECHMP). This overcomes the limitations of ECM and the other constituent process and allows exploiting their capabilities and advantages simultaneously with higher overall efficiency and productivity. The majority of material removal during an ECHMP is done by electrolytic and chemical dissolutions, while the role of the conventional machining/finishing action is basically limited to depassivate the electrochemically machined surface by removing non-conducting layers of metallic oxides and other compounds from the anode. This enhances the MRR and surface quality by changing the IEG conditions conducive for enhanced ED process.

2.1   Electrochemical Grinding (ECG)

2.1.1   Introduction

Grinding with diamond as abrasive is probably the only practical conventional process for finishing difficult-to-machine materials such as cemented carbides, high-strength-temperature-resistant (HSTR) alloys, and creep-resistant alloys. The scarcity of diamond and the challenges involved in diamond grinding wheel manufacture make this process extremely costly. ECG was developed by hybridizing ECM with the abrasive action of conventional grinding to machine hard and fragile electrically conducting materials efficiently, economically and productively without affecting the useful properties of these materials. ECG offers accurate and largely surface residual stress-free machining with no burrs and heat affected zone (HAZ) and, therefore, little distortion.

2.1.2   Equipment and Working Principle

Figure 2.1 depicts a typical set-up for an ECG process in which a metal-bonded abrasive grinding wheel acts as cathode and the workpiece as anode connected to a suitable DC power supply. The non-conducting abrasive particles protrude just beyond the surface of the bonding material of the wheel helping to maintain a constant inter-electrode gap and act as spacers. The grinding is performed as usual,

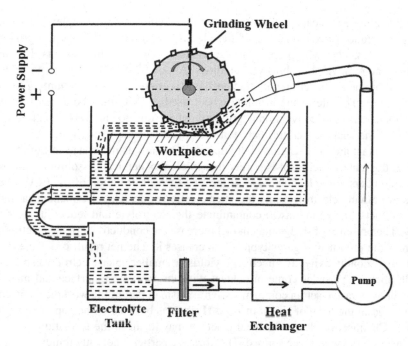

Fig. 2.1 Schematic of a typical set up for ECG process

but instead of coolant, a spray of an appropriate electrolyte is used. Bonding materials such as copper, brass, nickel or copper impregnated resin are commonly used for the manufacture of metal-bonded grinding wheels. The main functions of the abrasive particles in the ECG process are as follows: (i) to maintain the electrical insulation between anode workpiece and cathode grinding wheel and to maintain the effective IEG between them; (ii) to continuously remove any passive layer that may be formed on the workpiece surface by chemical reaction; and (iii) to determine workpiece shape and size. A commonly used abrasive is alumina (Al_2O_3) of mesh size 60–80. The main functions of the electrolyte used in the ECG process are (i) to produce the desired finish by ED; (ii) to conduct heat away from the inter-electrode gap; (iii) to flush reaction products away from the inter-electrode gap; and (iv) to minimize chemical wear of the conducting grinding wheel by maintaining its chemical inertness. Generally, sodium chloride and sodium nitrate are used as electrolytes in ECG [1–3].

2.1.3 Process Mechanism and Parameters

The electrochemistry of the ECG process implies that the electrochemical reactions (i.e. anodic dissolution, evolution of oxygen at anode and hydrogen at cathode and oxidation–reduction) occur at the workpiece material electrolyte boundary layer, while the other chemical reactions (i.e. chemical combinations, complex formations and precipitation) occur in bulk of the electrolyte solution [1]. As the grinding wheel rotates, the workpiece material is removed by simultaneous ED and mechanical removal by abrasive grinding. About 90–95 % of the material is removed by the ECM action with the mechanical grinding action accounting for the rest 5–10 %. The life of the grinding wheel used in ECG may be about 10 times more than that of a conventional grinding wheel due to the very small contact length (in the shape of an arc) between the wheel and the workpiece.

The contact arc in ECG can be divided into 3 zones as illustrated in Fig. 2.2. In **zone I**, the material removal is purely due to electrochemical dissolution. Material removal occurs at the leading edge of the ECG wheel. Rotation of the ECG wheel helps in drawing electrolyte into the inter-electrode gap. The electrochemical reaction products (including gases) contaminate the electrolyte and reduce its conductivity. The presence of sludge may in fact increase the conductivity of the electrolyte, whereas the presence of gases typically decreases it. The net result is a decrease in electrolyte conductivity. It effectively yields a smaller inter-electrode gap. As a result, abrasive particles come in contact with the workpiece surface and material removal by abrasive action commences. Thus, a small part of the workpiece material is removed in the form of chips. In **zone II**, the gas bubbles in the gap yield higher MRRs. Chemical or electrochemical reaction may result in the formation of a passive layer on the workpiece surface. The abrasive particles not only remove material from the work surface in the form of chips but also remove the non-reactive oxide layer. The removal of the non-reactive oxide layer is important as it promotes ED. In **zone III**, the material removal is done completely by electrochemical dissolution. This zone starts at the point where the wheel lifts just beyond the work surface. In this zone, pressure is released slowly. This zone contributes by removal of burrs that formed on the workpiece in **zone II** [2, 3].

The main process parameters that affect performance, efficiency and effectiveness of the ECG process are as follows: applied voltage and current density; type, concentration and delivery method of electrolyte; type, speed, pressure and

Fig. 2.2 Three zones of the contact arc in the ECG process

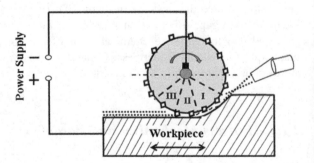

kinematic accuracy of the grinding wheel; and feed rate of workpiece. Selection of optimum values of these parameters helps to optimize surface finish and MRR.

2.1.4 Advantages, Limitations and Applications

Main advantages offered by ECG are as follows: negligible thermal damage to workpiece; limited grinding wheel wear; and no distortion of the workpiece. Metallographic examination of the surfaces finished by ECG revealed absence of structural changes, micro-cracks or any other defect [1]. Its main limitations are as follows: applicable for only electrically conductive work materials; not suitable for soft materials; requires tool dressing preparation for the grinding wheel; higher power consumption; corrosion problems of ancillary equipment due to the electrolyte; and high capital costs due to specialized grinding wheels.

ECG can routinely produce surface roughness values up to 0.1 μm. It has many industrial applications which include the following: grinding cemented carbide cutting tools, and thin-walled components; grinding of creep-resisting alloys (i.e. Inconel, Nimonic); grinding of titanium alloys, and metallic composites; burr-free sharpening by grinding of hypodermic needles; finishing of turbine blades made of superalloys; reprofiling of worn traction motor gears without affecting its hardness; form grinding of fragile aerospace honeycomb materials; and removal of fatigue cracks from steel structures used under seawater.

2.2 Electrochemical Honing (ECH) and Pulsed ECH

2.2.1 Introduction

Electrochemical honing is a hybrid fine finishing process combining the capabilities of *ECM* (capability to machine material of any hardness, production of stress-free surface with good finish and higher MRR) with the capabilities of *mechanical honing* (ability to correct shape/geometry-related errors, controlled generation of functional surfaces having cross-hatch lay patterns and compressive residual stresses) in a single process. At the same time, it overcomes some limitations of ECM along with certain limitations of mechanical honing (reduced tool life, low productivity due to frequent failure of honing sticks, inability to finish hardened workpiece and possibility of mechanical damage to the workpiece). ECH, therefore, provides a higher productivity alternative with many benefits that may produce surfaces that are not attainable by either of the processes when used individually [4].

2.2.2 Equipment and Working Principle

2.2.2.1 Finishing of Internal Cylinders by ECH

Figure 2.3a presents a schematic of the ECH set-up for finishing of an internal cylinder. It consists of five major subsystems: (i) DC power supply; (ii) tool for ECH process; (iii) the kinematic system for tool motion; (iv) electrolyte supply and cleaning system; and (v) machining chamber for holding and positioning workpiece. A power supply unit provides a DC voltage (3–40 V) and constant current (up to 200 A) across the electrolyte flooded inter-electrode gap. The positive terminal of the power supply is connected to the workpiece by means of a carbon-brush and slip ring assembly, while the negative terminal is directly connected to a brass ring mounted over the axle of the cathodic tool.

An ECH tool for finishing of cylindrical workpieces (shown in Fig. 2.3b) typically consists of a Teflon (PTFE) body over which a hollow stainless steel sleeve is placed having provision for an even number of equally spaced honing sticks to protrude out circumferentially by a light spring mechanism which can be used to adjust the required honing pressure. These honing sticks being electrically

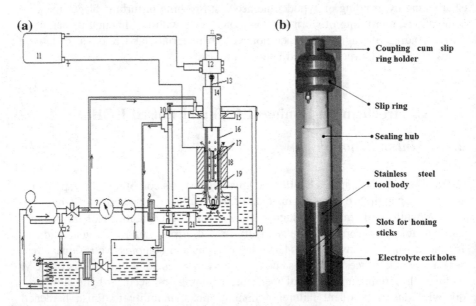

Fig. 2.3 a Schematic diagram of an ECH set-up; and **b** photograph of the tool used for finishing internal cylinders by ECH [5], with kind permission from Springer. *1* Electrolyte settling tank; *2* Flow control valve; *3* 1st stage filter cum magnetic separator; *4* Electrolyte storage tank; *5* Temperature control system; *6* Stainless steel electrolyte supply pump; *7* Pressure gauge; *8* Flow meter; *9* 2nd stage filter cum magnetic separator; *10* Mist collector; *11* DC power source; *12* Carbon brush and slip ring assembly; *13* Copper connector; *14* Seal hub; *15* Hydraulic cum mechanical seal; *16* Tool body; *17* Honing sticks; *18* Workpiece; *19* Electrolyte exit holes; *20* Work chamber; *21* Fixture cum electrolyte inlet sleeve

non-conducting maintain a uniform inter-electrode gap and preferentially remove the non-conductive passive layer of metal oxide from the high spots to correct errors/deviations in geometry/shape of the cylindrical workpiece. The tool is provided a precisely controlled combination of rotation and reciprocation simultaneously. Rotary motion is provided by a speed-controlled DC servo motor, while the reciprocating motion is provided by a microprocessor-controlled stepper motor.

2.2.2.2 Finishing of Gears by ECH

The equipment for gear finishing by ECH has all subsystems same that used in finishing of internal cylinders, except the machining chamber, which is significantly different for gear finishing by ECH. Figure 2.4a, b shows photograph of the machining chamber for high-quality finishing of spur gear developed by Naik et al. [6] and for helical gears by Misra et al. [7].

Gear finishing by ECH requires that the anodic workpiece gear meshes with a specially designed cathode gear for the ECM action to occur while simultaneously meshing with a honing gear so that the mechanical honing can also take place. All three gears should have the same involute profile and module. The honing gear is a helical gear and can be either an abrasive impregnated gear or manufactured from a material substantially harder than the workpiece gear. It is mounted on a floating stock to ensure dual flank contact between the honing and workpiece gear. For *finishing spur gears* by ECH, the honing gear is mounted in cross-axis arrangement with the workpiece gear to reduce the tooth surface contact and therefore pressure required for finishing (Fig. 2.4a).

Helical gear finishing by ECH does not require cross-axis arrangement because the honing and cathode gears have opposite helix angles than the workpiece gear. Therefore, if the workpiece gear is a right-handed helical gear, then the cathode and honing gears will be left-handed and vice versa (refer to Fig. 2.4b).

The cathode gear is to be designed and fabricated such that while meshed with the electrically conducting anodic workpiece gear, no short circuiting should occur while maintaining the required inter-electrode gap necessary for the ED of the workpiece gear. For the cylindrical gears, this is ensured by sandwiching a conducting layer between two non-conducting layers and undercutting the conducting layer by a distance equal to the inter-electrode gap as compared to the non-conducting layers. The axes of shafts of the workpiece and cathode gears are parallel to each other for finishing the spur and helical gears (Fig. 2.4a, b). The rotation of the workpiece gear is controllable by means of a DC motor, while cathode and honing gears rotate by virtue of their engagement with the workpiece gear. Since the entire face width of the cathode gear is not electrically conducting, a reciprocating motion is also required for the workpiece gear. This is implemented by a controlled stepper or servo motor, therefore ensuring finishing of the entire face width of the workpiece gear.

(a)

(b)

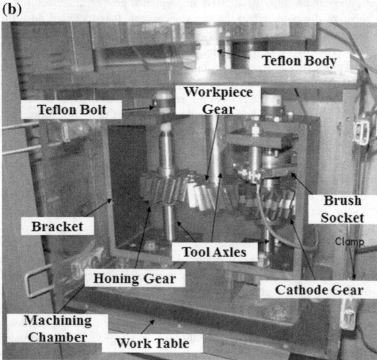

Fig. 2.4 Photographs of machining chamber for high-quality finishing of **a** spur gear [6] and **b** helical gear [7] by ECH process, with kind permission from Sage

Finishing of *bevel gears* by ECH is much more difficult than cylindrical (i.e. spur or helical) gears due to their complex geometry which inhibits reciprocation of the workpiece gear that is required for finishing the entire face width. This problem was solved by Shaikh et al. [8] that envisaging a novel concept of a *twin complementary cathode gear* arranged as shown in Fig. 2.5a. In this arrangement, one of the cathode gears '4' has a conducting layer of copper sandwiched between two insulating layers of metalon, while the other complimentary cathode gears '3' have an insulating layer of metalon sandwiched between two conducting layers of copper. The workpiece gear '1', which acts as the anode, is mounted on the spindle of a bench drilling machine. To avoid short circuiting, an inter-electrode gap is provided between the cathode and anode gears by undercutting the conducting layer equal to the required distance of the inter-electrode gap as compared to the insulating layers. A honing gear '2' is mounted to the rear of the workpiece gear. The workpiece gear is actively rotated, thereby ensuring rotation of both the honing and cathode gears due to the inter-meshing. The surface contact established because of the close meshing between the honing and workpiece gears facilitates the removal of the passivating layer and consequently exposes fresh workpiece surface for further finishing by the electrochemical action [8]. A photograph of the machining chamber based on this principle is shown in Fig. 2.5b. The finishing occurs as a result of the ED that takes place when the required quality of an appropriate electrolyte (generally mixture of NaCl and NaNO$_3$) is available at the inter-electrode gap at the appropriate temperature and pressure along with a suitable DC current between the anode and cathode gears.

During the process of material removal from the tooth flank, the electrolyte forms a passivating metal oxide layer on the workpiece gear tooth surface which inhibits further finishing by the electrolysis action. This layer is abraded by the honing gear. Surface peaks both along the tooth face and along the involute profile have minimum thickness of the oxide layer; therefore, they are exposed early for further finishing by electrolysis action once again. As the process carries on, the geometric accuracy of the workpiece gear is rapidly improved.

Pathak et al. [10] showed that the use of a pulsed power supply during ECH (a process referred to as pulse-ECH or PECH), simultaneously improves surface finish and micro-geometry of the bevel gear by a significant margin, thus enhancing their service life and operating performance. They achieved more than 50 % improvement in location errors (i.e. pitch error and run-out) in PECH-finished bevel gears as compared to ECH-finished bevel gears as reported by Shaikh et al. [8]. Misra et al. [11] used PECH for improving surface finish of spur gears and reported that gravimetric electrolyte composition of 75 % sodium chloride (NaCl) and 25 % sodium nitrate (NaNO$_3$) with an electrolyte temperature of 30 °C yielded the best results. This proves superiority of PECH over the ECH process for gear finishing applications. However, the use of PECH may take longer time to achieve the desired quality of the gears, making this process less productive than ECH.

(a)

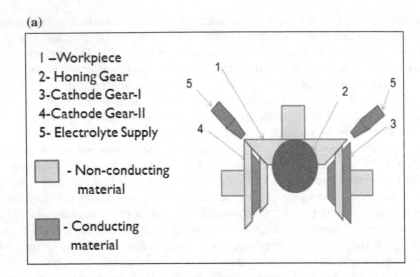

(b)

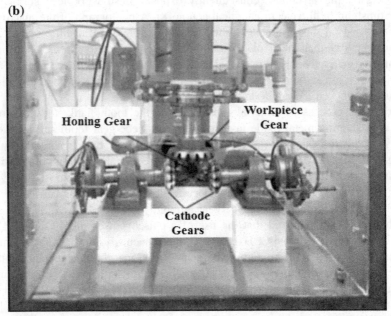

Fig. 2.5 a Concept of twin complementary cathode gears for high-quality finishing of bevel gears by ECH; **b** photograph of the machining chamber based on this concept [9], with kind permission from Elsevier

2.2.3 Process Parameters and Mechanism

Parameters related to the power supply, electrolyte, honing tool and workpiece affect performance, efficiency and effectiveness of the ECH process. These parameters include the following: type of power supply (continuous or pulsed); applied voltage and current; type, composition, concentration, flow rate, temperature and pressure of the electrolyte; type and size of abrasives used in the honing tool; honing tool hardness; honing pressure; rotary and/or reciprocating speed of the honing tool (for internal cylinders) or workpiece (for gears); and electrochemical properties of the workpiece.

Mechanism of finishing the gears by ECH explained by Shaikh and Jain [9] involves cyclic sequence of finishing by ED and mechanical honing which improves the geometric accuracy and surface finish of the workpiece gear. More than 90 % of the material is removed by ED. As far as the surface topography is concerned initially, the distances between peaks on the workpiece and cathode surfaces are less than the corresponding distances between valleys (Fig. 2.6a). Consequently, more material is removed from the peaks of the workpiece surface as compared to its valleys by ED, thus truncating the highest peaks and giving rapid improvement in the workpiece surface finish. At the end of the ED process, a passivating metal oxide layer is formed on the workpiece surface (Fig. 2.6b, d and f), which is removed by the honing action so that further ED can continue. At the end of honing action, certain valleys may still be covered with the passivating layer as shown in Fig. 2.6c, e and g.

In the next cycle of ED, these peaks will be exposed for further smoothing. Consequently, these peaks are truncated along with material above the valley giving a smoother surface as shown in Fig. 2.6g. As the ECH process continues, surface finish and geometric accuracy initially improve rapidly but follow the law of

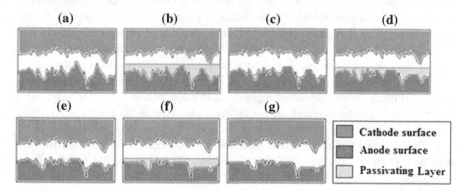

Fig. 2.6 Sequential finishing by the ECH process depicting the surface profiles of the anode and cathode. **a** Initial surface before ECH; **b** after first phase of electrolytic dissolution; **c** after first phase of honing action; **d** after second phase of electrolytic dissolution; **e** after second phase of honing action; **f** after third phase of electrolytic dissolution; and **g** after third phase of honing action [9], with kind permission from Elsevier

diminishing return due to increasing inter-electrode gap between the surfaces of the cathode and workpiece gears and consequent decreasing MRR. Therefore, whenever the desired surface finish or geometric accuracy or both are achieved, the process can be stopped.

Based on the above-mentioned mechanism of material removal, Shaikh and Jain [9] developed models of MRR and the arithmetic mean of maximum peak-to-valley heights (R_z) for the flank surface of bevel gears finished by the ECH process as a function of rotary speed of the workpiece gear and the inter-electrode gap current. The model also indirectly took into account the effects of other electrolytic parameters such as concentration, temperature and flow rate. The contribution of ECD in the MRR and surface roughness was modelled using Faraday's law of electrolysis, while the contribution of the mechanical honing was modelled considering material removal as a process of uniform wear [9].

2.2.4 Advantages, Limitations and Applications

ECH offers many useful advantages as follows: (i) material of any hardness (but electrically conducting) can be finished by ECH; (ii) it produces surfaces with a distinct cross-hatch lay pattern that is beneficial for oil retention; (iii) ECH not only produces high-quality surface finish and surface integrity but also has the ability to correct errors/deviations in geometry/shape such as out of roundness or circularity, taper, bell-mouth hole, barrel-shaped hole, axial distortion and boring tool marks for cylindrical surfaces and the ability to correct form errors (i.e. deviations in lead and profile) and location errors (i.e. pitch deviations and run-out) for cylindrical and conical gears; (iv) it is faster when compared to ECM and mechanical honing. ECH can finish materials up to 5–10 times faster than mechanical honing and four times faster than internal grinding. The benefit is more pronounced for higher material hardness; (v) low heat generation thus making it suitable for the processing of parts that are susceptible to heat distortions; (vi) increased life of abrasive sticks/tool due to the limited contribution of mechanical honing to the process; and (vii) low working pressure implies less distortion while finishing thin-walled sections.

Despite the numerous advantages, ECH does exhibit some limitations as follows: (i) it can be used for finishing electrically conductive materials only; (ii) it is more costly than the mechanical honing due to the cost of the electrical and fluid handling elements, need for corrosion protection, costly tooling and longer set-up time. This makes ECH more economical for longer production runs than for tool room and job-shop conditions; (iii) it cannot finish blind holes easily; and (iv) ECH cannot correct location of hole or perpendicularity [4].

Materials that can be finished by ECH include cast tool steels, high-alloy steels, carbide, titanium alloys, Incoloy 901, 17-7PH stainless steel, Inconel and gun steels. ECH is an ideal choice for superfinishing, improving the surface integrity and increasing the service life of the critical components such as internal cylinders, transmission gears, carbide bushings and sleeves, rollers, petrochemical reactors, moulds and dies, gun barrels and pressure vessels, which are made of very hard and/or tough wear-resistant materials, most of which are susceptible to thermal distortions. Therefore, ECH is widely used in the automobile industry, aerospace, petrochemical industry, power generation and fluid power industries. It has been successfully used for finishing bore sizes ranging from 9.5 to 300 mm and length up to 600 mm [4]. ECH can achieve surface roughness up to 50 nm and tolerances of ±0.002 mm [5].

2.3 Electrochemical Buffing (ECB)

2.3.1 Introduction

Electrochemical buffing (ECB) is a non-contact-type electrochemical hybrid machining process (ECHMP), combining process principles and advantages of electrochemical finishing (ECF) and the conventional buffing process. The most significant challenges of the conventional buffing process are the dependence of surface quality on operator's skills, residual stresses, micro-cracks and control of buffing pressure, i.e. pressure with which the buff is pressed against the workpiece. The most significant limitations of the ECF process are as follows: formation of a passivating metallic oxide layer at the anode (workpiece) which prohibits further finishing by the electrolytic process and poor surface finish due to selective dissolution, sporadic breakdown of the anodic film, flow separation and formation of eddies and evolution of hydrogen gas at the cathode. Hybridization of ECF with conventional buffing in ECB process overcomes most of these limitations by exploiting the individual capabilities and advantages in a single process simultaneously.

2.3.2 Equipment and Process Mechanism

Figure 2.7 depicts the working principle of the ECB process in which a rotating buffing pad impregnated with fine abrasive particles is pressed against the workpiece in the presence of an electrolyte. The buffing pad acts as the cathode, while the workpiece becomes the anode. A suitable DC voltage is applied in the presence of an appropriate electrolyte usually sodium nitrate or sodium chloride. The polishing performance of ECB is not affected by the electrolyte temperature because the ECF action acts to assist conventional buffing only. This eliminates need of a heat exchanger [12].

Electrolyte used in the ECB process is generally 20 wt% of aqueous solution of sodium nitrate which is quite environment-compatible. Sodium nitrate solutions do not deteriorate with time and therefore lends itself to long-term frequent use. Harmless and safe nature of sodium nitrate solution makes the polishing action by

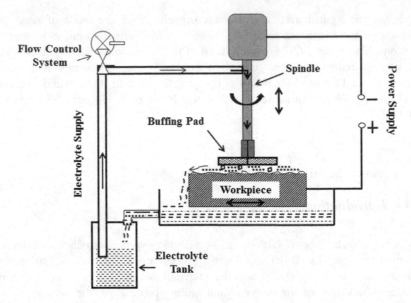

Fig. 2.7 Working principle of ECB process

ECB simpler because it does not require strong-acid compatible plumbing and other components such as flanges and gaskets, unlike electropolishing (EP) process which uses harmful mixture of hydrofluoric and sulphuric acid requiring carefully controlled handling of it and many additional facilities [12]. ECB not only offers extended electrolyte service life, but also simple operational procedures and affordable liquid waste handling. ECB may therefore be considered as one of the green hybrid machining process.

The key process control parameters that affect the performance of ECB process are electrolyte flow rate, type and size of abrasive particles, rotary speed of the buffing pad, buffing pressure, buffing duration, DC voltage and current density.

2.3.3 Applications

ECB process may be used with numerous electrically conducting materials including steel, stainless steel, aluminium, copper, titanium and molybdenum. ECB can buff ultra-high purity components to a mirror finish without any residual stresses and micro-cracks. ECB can remove altered surfaces layers and achieve ultra-flat surfaces without contaminants and micro-dust particles. The ECB process can be used for almost any size part or component from thin tubes of a few millimetres in diameter to large tanks producing buffed surfaces with a maximum surface roughness of 100 nm. Due to this flexibility, Kato et al. [12] demonstrated ECB on complicated structures including niobium cavity cells and endgroups.

2.4 Electrochemical Deburring (ECDe)

2.4.1 Introduction

Electrochemical deburring (ECDe) is a process of removing 'burrs' from the man-ufactured components especially from and at the areas which are difficult to access with any other deburring process. A burr is a small three-dimensional sharp pro-jection protruding from a manufactured and/or finished surface. There are many types of burrs produced by different manufacturing and finishing processes, namely compressive burr, corner burr, edge burr, entrance burr, exit burr, feather burr, flash burr, hanging burr, parting-off burrs, roll over burr and tear burr [3]. The presence of burrs may affect functional aspects (i.e. assembly, handling, positioning, mounting, wear and breakdown), physiological aspects (injury during further processing, assembling, use, maintenance, repair and utility, etc.) and aesthetic aspects (im-pression, sale, advertising, etc.) of a product. Precise control of the manufacturing process parameters and process conditions can reduce the occurrence of burrs to a certain extent, but demand for high-quality finishing and appearance of products necessitates complete removal of all types of burrs before a product is marketed.

The main deburring processes can be classified into five categories [3]: (i) **mechanical deburring** which minimizes the burrs using brushes, scrappers, cutting tools, etc., in processes such as hand deburring, power brushing and scrapping; (ii) **abrasive deburring** uses abrasives in different forms (loose agitated, jet, putty) to remove the burrs in processes such as tumbling, barrel finishing, spindle fin-ishing, vibratory deburring, sand blasting, magnetic loose deburring, abrasive flow machining (AFM), and abrasive water jet deburring, using loose abrasives, semi-solid putty and liquid abrasive flow; (iii) **thermal deburring** is a process where burrs are burnt away using some heat source for a short duration in processes such as the thermal energy method, flame melting, resistance heating and hot wire; (iv) **chemical deburring** dissolves burrs in a chemical medium in processes such as chemical barrel finishing, chemical spindle finishing, chemical vibratory finishing, chemical magnetic loose brush, etc.; and (v) **electrochemical deburring (ECDe)** removes burrs by electrolytic dissolution.

Among the different types of deburring processes, ECDe is used for remotely located and inaccessible areas where other deburring processes are not effective.

2.4.2 Process Details

Figure 2.8 shows a typical arrangement for ECDe of holes in which the part to be deburred is made the anode and is placed in a fixture, which positions a specially designed cathode tool in close proximity to the burrs to be removed. An appropriate electrolyte is supplied at a suitable pressure (0.3–0.5 MPa) to the gap between the cathodic deburring tool and the burrs. When a suitable DC current (50–500 A) is

Fig. 2.8 Typical arrangement for ECDe of a hole [3]

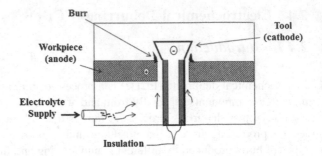

supplied, the burr dissolves forming a controlled radius. The cathode tool has as much working areas as practical so that several interfaces are deburred at any given time. The tool surfaces at interfaces where deburring is not required are insulated. The deburring tool should also have a similar contour as that of the workpiece, thus maintaining an inter-electrode gap in the range of 0.1–0.3 mm. The tool tip should overlap the machined area by 1.5–2 mm in order to produce a proper radius. The use of a rotating and feeding tool electrode enhances the deburring process by creating turbulent flow in the inter-electrode gap. The spindle rotation is reversed to increase the electrolyte turbulence. Selecting a suitable electrolyte plays an important role in the ECDe process. Commonly used electrolytes are aqueous solutions of either sodium chloride or sodium nitrate [3].

2.4.3   Advantages and Applications

The main advantages of ECDe are as follows: (i) replacement of costly hand deburring process; (ii) increased product quality and reliability; (iii) removal of burrs at the required accuracy, uniformity, radius, and edge quality; (iv) reduced personnel and labour cost; and (v) possibility of automating for higher productivity.

ECDe can be used to deburr gears, spline shafts, milled components, drilled holes and punched blanks. The process is particularly efficient for hydraulic system components such as spools and sleeves of fluid distributors. Normal cycle time for ECDe is between 30 and 45 s. ECDe can finish a burr of 0.5 mm height to a radius of 0.05–0.2 mm with a surface roughness up to 2–4 μm [13].

2.5   Electrochemical Superfinishing (ECSF)

2.5.1   Introduction

Conventional superfinishing process removes surface micro-irregularities with slowly and continuously reciprocating and oscillating abrasive sticks along the length of a rotating workpiece. It may however retain surface micro-irregularities such as waviness and out of roundness. Electrochemical superfinishing (ECSF) came into existence specifically to overcome the limitations of conventional superfinishing. ECSF combines the advantages of electrolytic dissolution associated with ECM process and the mechanical abrasion action of conventional superfinishing overcoming their individual limitations.

2.5.2   Equipment and Working Principle

In ECSF, the ED is assisted with the mechanical abrasion from a separate cathodic tool electrode or a metal-bonded diamond abrasive stick (Fig. 2.9) for finishing. The workpiece is made of the anode by connecting it to the positive terminal of a DC power supply, while the tool or a metal-bonded diamond stick is wired as the cathode. The workpiece is usually rotated, while a combination of reciprocating and oscillating motion is induced at the cathodic tool. The abrasive particles protruding from the stick maintain the required inter-electrode gap between the anodic workpiece and cathodic tool. Light stick pressure is used to avoid metallurgical

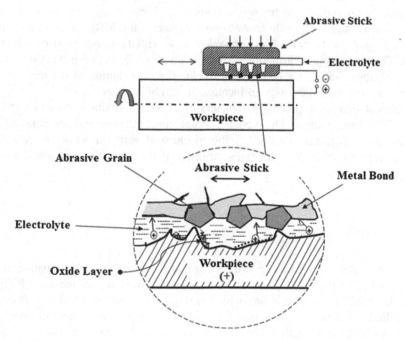

Fig. 2.9 Working principle and mechanism of material removal of ECSF process using metal-bonded diamond abrasive stick

damage due to a scrubbing motion by the abrasive sticks and to produce a superfine surface.

2.5.3 Process Mechanism and Parameters

Similar to the ECH process, the ED in ECSF process is also accompanied by the formation of a passive metal oxide film on the anode surface. Initially, the abrasive particles scrub away the surface irregularities protruding from the ideal surface (as shown in Fig. 2.9). During the next ED phase, these irregularities are subjected to an increased ED as compared to those areas still covered with the protective oxide film. Under such circumstances, the protecting metal oxide film can be used to correct the geometric inaccuracies such as cylindricity and roundness errors. Tolerances of about ±0.013 mm on diameter and a roundness and straightness of less than 0.007 mm can be achieved with this process [13].

Datta and Landolt [14] observed high instantaneous current densities on the use of pulsed DC voltage. This is possible because each current pulse is followed by a relaxation phase of zero current, which allows for the removal of reaction products and the heat generated by the Joule effect from the inter-electrode gap. Hofy [15] investigated a linear increment in MRR with current density during a series of experiments on pulsed electrochemical superfinishing. A rise in the scrubbing speed, voltage and duty cycle leads to an increase in the MRR. The high energy available enhances the oxide film removal process with a consequent rise in the ED phase. The percentage contribution of the ED phase varies between 0 at 20 % duty cycle and about 95 % at a 100 % duty cycle. The contribution of the mechanical abrasion actions increases with an increase in scrubbing speed.

The most significant process parameters of ECSF are similar to those identified for ECM. These include DC voltage, type; concentration and temperature of electrolyte, the parameters related to the mechanical abrasion action, namely frequency and amplitude of oscillations; abrasive grain characteristics and abrasive stick pressure.

2.5.4 Applications

The higher material removal capabilities combined with its ability to finish to close tolerances enable the ECSF process to be used in a wide range of industry. It eliminates the need for initial grinding which is required before conventional superfinishing. ECSF can be used to produce required dimensions in difficult-to-machine materials particularly to those parts that are susceptible to heat and distortion. The ECSF process effectively eliminates the thermal distortion problem (normally found in conventional superfinishing) as the majority of material removal occurs electrochemically in an electrolyte-cooled atmosphere. This also enables it to produce burr-free components. Hofy [15] reported reduction of roundness error from 24 to 8 μm and average surface roughness value between 2.25 and 0.65 μm after ECSF for 2 min at 19 V, 67 % duty cycle and a scrubbing speed of 18.55 m/min.

2.6   Electrochemical-Type Advanced Drilling Processes

Electrochemical-type advanced drilling processes (ADPs), which utilize an acidic electrolyte include electrochemical drilling (ECD), capillary drilling (CD), shaped tube electrodrilling (STED), electrolytic jet drilling (EJD) and jet electrolyte drilling (JED). Among various advanced hole drilling processes, electrochemical-type ADPs meet the various requirements of micro/small deep hole drilling (i.e. hole diameter less than 1 mm with aspect ratio, i.e. ratio of hole depth to hole diameter, more than 20) with high productivity and better hole geometry and resulting in minimum surface damage to the workpiece material. The main advantages of these processes include better surface finish, absence of residual stresses, no tool wear, burr-free and distortion-free holes and simultaneous drilling of large number of holes [16]. Table 2.1 presents details of important process parameters and capabilities as relevant for different electrochemical-type ADPs. The basic working principles for the different processes are depicted in Fig. 2.10.

2.6.1   Electrochemical Drilling (ECD)

ECD is a controlled electrolytic drilling process into an electrically conducting workpiece material, operating as the anode, utilizing a tubular shaped tool (preferably made of brass, copper or stainless steel). The outer surface of the tool is insulated except at the tip where machining occurs and where the required inter-electrode gap is maintained (Fig. 2.10). Hole drilling occurs in the anodic workpiece material by ED by applying a suitable DC current across the inter-electrode gap in the presence of an appropriate electrolyte (an aqueous solution of salts such as NaCl, $NaNO_3$ or their mixture) that is pumped through the tubular tool at the required pressure and temperature. The electrolyte also flushes the reaction products away while also removing the heat generated during the process to obtain and maintain an improved MRR. The ECD process has two significant limitations, i.e. tool insulation loss and stray removal. Tool insulation loss in ECD occurs mainly due to clogging of the drilled holes by the salt-based electrolytes. Stray removal implies the removal of materials from the side wall of the hole. Zhu and Xu have attempted to address these limitations by using good-quality insulation and/or using a dual pole tool which uses a metallic bush-insulated coating [17]. This improves the machining accuracy and process stability.

Table 2.1 Comparison of process parameters and capabilities of electrochemical-type advanced drilling processes [16]

Parameter	Electrochemical drilling (ECD)	Shaped tube electro drilling (STED)	Capillary drilling (CD)	Electrolytic jet drilling (EJD) or electrolytic stream drilling (ESD) Penetration	Dwell	Jet electrolyte drilling (JED)
Type of electrolyte	NaCl, NaNO$_3$	HNO$_3$, H$_2$SO$_4$	HNO$_3$, H$_2$SO$_4$, HCl	HNO$_3$, H$_2$SO$_4$, HCl		HNO$_3$, H$_2$SO$_4$
Electrolyte pressure (bar)	3–10	3–10	3–20	3–10		10–60
Tool	Tube-shaped tool made of brass, copper and stainless steel	Titanium tube	Glass capillary with gold, platinum or titanium wire	Glass nozzle with capillary end with gold, platinum or titanium wire		Platinum nozzle
Tool feed (mm/min.)	Equal to linear MRR	1.0–3.5	1.0–4.0	1.0–3.5	No tool feed	No tool feed
Operating voltage (V)	10–30	5–15	100–200	150–850	150–850	400–800
Hole diameter (mm)						
Minimum	1.0	0.65	0.2	0.125	0.125	0.125
Maximum	7.5	6.35	0.5	1.0	1.0	1.0
Aspect ratio (ratio of hole depth to hole diameter)						
Typical	8:1	16:1	16:1	16:1	–	16:1
Maximum	20:1	300:1	100:1	40:1	10:1	30:1
Hole depth (mm)	125	125	–	19	5	–
Avg. surface roughness R_a (µm)	0.3–1.6	0.8–3.2	–	0.25–1.6	0.25–1.6	–

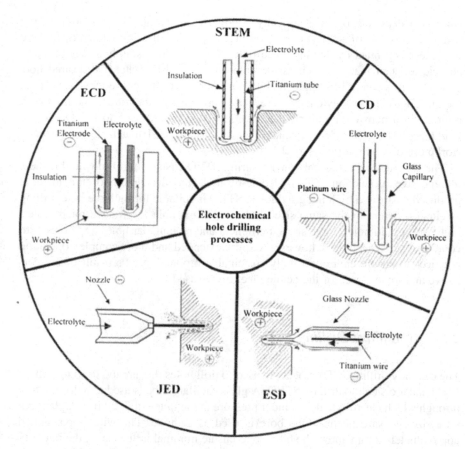

Fig. 2.10 Process principles of various electrochemical-type advanced drilling processes [16], with kind permission from Elsevier

2.6.2   *Shaped Tube Electrodrilling (STED)*

Drilling deep holes with aspect ratio in excess of 20 are challenging mainly due to the difficulty in removal of the machined material as the hole is being drilled. STED is a special process developed by *General Electric Co.* for drilling very high aspect ratio hole (up to 300), which cannot be drilled by any conventional process or ECD (due to formation of insoluble precipitates). STED is a modified version of ECD process that utilizes an acidic electrolyte which dissolves the material removed instead of forming a precipitate. Aqueous solution of acidic electrolytes such as sulphuric acid, nitric acid and hydrochloric acid with 10–25 % concentration is used in this process. A 5–15 V DC potential difference is applied, which is in comparison with ECD (10–30 V), slightly lower. The lower voltage requirement is primarily due to the use of more conductive acid electrolytes instead of the mainly neutral electrolytes as used in the ECD process. The STED process drills holes by controlled deplating of an electrically conductive material. The deplating action takes place as part of an electrolytic cell made up of a cathodic tool and the anodic workpiece separated by flowing electrolyte. The cathode is simply a metal tube of an acid-resistant material such as titanium and shaped to match the desired hole geometry. It is carefully straightened and insulated over the entire length except at the tip. The acid electrolyte is fed under pressure through the tube to the tip, and it returns via a narrow gap along the outside of the coated tube to the top of the workpiece. The electrode is displaced at a constant feed matching the effective workpiece dissolution rate (Fig. 2.10).

The absence of mechanical contact during STED ensures uniform wall thickness in repetitive production. The molecule-by-molecule dissolution of the material produced unstressed high integrity holes. STED is suitable for multiple hole drilling of either different or the same sized holes. STED is able to produce micro-holes, high aspect ratio holes, large shaped elliptical and rectangular holes and holes with contoured surfaces. It does, however, require extended and more complex operating practices to ensure environmentally sustainable production as a result of the corrosive and toxic nature of the acidic electrolytes used.

2.6.3 Capillary Drilling (CD)

The capillary drilling (CD) process is used to drill holes that are too deep to drill by EDM and too small to drill by STED. A glass capillary tube is used as a drilling tool through which electrolyte flows under pressure in the range of 3–20 bar. A platinum wire sized to suit the fine tube bore is used as cathode. The wire is positioned approximately 2 mm above the tube tip to ensure minimal influence on the integrity and the direction of electrolyte flow at the tip. Higher DC voltages in the range of 100–200 V are used to overcome the resistive path of current flow due to longer electrolyte flow path. It has been successfully used for drilling trailing edge holes (dia. 0.2–0.5 mm, depth 8–16 mm) in high-pressure gas turbine blades. The process finds a wide range of applications for drilling holes in production components with positioning and diametral tolerance values up to ±0.05 mm [16].

2.6.4 Electrolytic Stream or Jet Drilling (ESD or EJD)

Electrolytic stream or jet drilling (ESD or EJD) is an efficient electrochemical-type ADP for drilling micro to small holes in any electrically conductive material without affecting its properties. In EJD, a negatively charged pressurized jet of acidic electrolyte exits through a finely drawn glass tube nozzle to impinge on the anodic workpiece, thereby achieving controlled ED. Nozzles used in EJD are made of glass tube which is drawn to a small diameter, thus forming a capillary at one end. A dilute aqueous solution of either sulphuric acid (H_2SO_4) or hydrochloric acid

(HCl) is used as electrolyte. HCl is preferred for drilling holes in corrosion-resistant materials such as aluminium and titanium, while H_2SO_4 is the preferred electrolyte for drilling holes in carbon steel, cobalt alloys, nickel alloys, chromium alloys, superalloys and stainless steel. The electrolyte exiting the nozzle is negatively charged either by using a small titanium/platinum/gold wire placed inside the large diameter section of the nozzle or by using a metallic sleeve. A suitable voltage is applied across the two electrodes. Material removal occurs through ED when the electrolyte jet strikes the workpiece [3]. A much longer and thinner electrolyte flow path requires a much higher voltage (150–800 V) to obtain sufficient current flow. The minimum diameter hole is strongly influenced by the nozzle diameter, electrolyte pressure and overcut [2]. A gap of 2–4 mm (known as standoff distance, SOD) must be maintained between the two electrodes.

There are two variants of EJD: *dwell EJD* in which the nozzle is not actively fed into the workpiece and *penetration EJD* in which the nozzle is displaced at a finite feed. In dwell EJD, the nozzle tip is fixed at a predetermined distance from the work surface and drilling is therefore purely done by the electrolyte jet. This does however limit the depth of the hole to be drilled and the maximum accuracy that can be obtained. Dwell EJD is used for drilling shallow micro-holes and in circumstances where the workpiece configuration or machine capabilities do not permit movement of the nozzle. In *penetration EJD*, the nozzle is fed towards the workpiece with a finite feed rate to maintain a constant IEG. Penetration EJD can achieve aspect ratios of up to 40:1, whereas dwell EJD can only achieve aspect ratios up to 10:1.

References

1. Bhattacharyya A (1973) New technology. Institution of Engineers (India), Calcutta
2. Bennedict GF (1987) Nontraditional manufacturing processes. Marcel Dekker Inc., New York
3. Jain VK (2002) Advanced machining processes. Allied Publishers Pvt Ltd., New Delhi
4. Jain NK, Naik LR, Dubey AK, Shan HS (2009) State of art review of electrochemical honing of internal cylinders and gears. Proc IMechE Part B: J Eng Manuf 223(6):665–681
5. Dubey AK, Shan HS, Jain NK (2008) Analysis of surface roughness and out-of-roundness in electro-chemical honing of internal cylinders. Int J Adv Manuf Technol 38(5–6):491–500
6. Naik LR, Jain NK, Sharma AK (2008) Investigation on precision finishing of spur gears by electrochemical honing. In: Shunmugam MS, Babu NR (eds) Proceedings of 2nd international and 23rd AIMTDR conference, IIT Madras, India, 15–17 Dec 2008, pp 509–514
7. Misra JP, Jain NK, Jain PK (2010) Investigations on precision finishing of helical gears by electro chemical honing (ECH) process. Proc IMechE Part B: J Eng Manuf 224(12): 1817–1830
8. Shaikh JH, Jain NK, Venkatesh VC (2013) Precision finishing of bevel gears by electrochemical honing. Mater Manuf Processes 28(10):1117–1123
9. Shaikh JH, Jain NK (2014) Modeling of material removal rate and surface roughness in electrochemical honing of bevel gears. J Mater Process Technol 214(2):200–209
10. Pathak S, Jain NK, Palani IA (2015) Process performance comparison of ECH and PECH for quality enhancement of bevel gears. Mater Manuf Processes 30(7):836–841

11. Misra JP, Jain PK, Jain NK, Singh H (2012) Effects of electrolyte composition and temperature on precision finishing of spur gears by pulse electrochemical honing (PECH). Int J Precis Technol 3(1):37–50
12. Kato S, Nishiwaki M, Tyagi P V, Azuma S, Yamamoto F (2010) Application of electro chemical buffing onto Niobium SRF cavity surface. In: Proceedings of IPAC10, Kyoto, Japan, pp 2929–2931
13. Hofy HE (2005) Advanced machining processes: non-traditional and hybrid machining processes. McGraw Hill Inc., New York
14. Datta M, Landolt D (1983) Electrochemical saw using pulsating voltage. J Appl Electrochem 13:795–801
15. Hofy HE (1990) Characteristics of pulsed EC-superfinishing. Alexandria Eng J 29(1):83–100
16. Sen M, Shan HS (2005) A review of electrochemical macro-to-micro hole drilling processes. Int J Mach Tools Manuf 45(2):137–152
17. Zhu D, Xu HY (2002) Improvement of electrochemical machining accuracy by using dual pole tool. J Mater Process Technol 129:15–18

Chapter 3
Thermal Hybrid Machining Processes

Abstract This chapter describes thermal-type HMPs in which EDM as the primary process of material removal is combined with either a conventional machining/finishing process or with an electrolytic dissolution-based process for improved machining/finishing characteristics and workpiece surface integrity. This hybridization enables thermal HMPs to reduce processing times by 2 to 3 times when compared to their constituent processes while producing surface finishes up to 0.1 μm. Some of the important applications of thermal HMPs are as follows: drilling of micro-holes and the production of slits in quartz and glass, machining of metal matrix composites and ceramics, and dressing of grinding wheels.

Keywords Abrasion · Diamond · Discharge · EDM · Electrochemical · Electrolysis · Gas film · Grinding · Pyrex

EDM is the primary process utilized in thermal-type HMPs, which implies that workpiece material is removed by melting and evaporation caused by a controlled series of high-energy electrical discharges (sparks). This mechanism can be combined either with an electrolytic dissolution-based process or with a conventional machining/finishing mechanical abrasion process to achieve higher material removal rates and improved product quality. These processes are presented and discussed in the sections that follow.

3.1 EDM Combined with Conventional Machining

3.1.1 Electric Discharge Grinding (EDG)

3.1.1.1 Introduction

EDG is a thermal HMP comprising of EDM as a thermoelectric process combined with the abrasion characteristics of grinding that was especially developed for machining/finishing difficult-to-machine hard and brittle electrically conductive

© The Author(s) 2016
K. Gupta et al., *Hybrid Machining Processes*,
Manufacturing and Surface Engineering,
DOI 10.1007/978-3-319-25922-2_3

materials [1–3]. In this process, an electrically conductive grinding wheel is used as a tool electrode instead of stationary tool electrode as used in EDM. Material removal occurs due to the electrodischarge action, while the abrasive component of the process ensures effective flushing which results in improved material removal rate and enhanced surface finish as compared to the conventional EDM process. When compared to EDM, EDG is the more productive process especially when machining extremely hard materials where improvements of between 2 and 3 times are possible when compared to conventional grinding. The absence of physical contact between workpiece and wheel makes the machining of thin and fragile specimen made of electrically conductive materials possible.

3.1.1.2 Equipment and Working Principle

The setup of a typical EDG machine is illustrated in Fig. 3.1. It consists of a tool electrode (cathode) in the form of a grinding wheel made of an electrically conductive material (such as graphite, copper or brass) but without any embedded abrasive, workpiece (anode), dielectric tank, servo system and a power supply system. The grinding wheel (usually made of graphite) rotates at an appropriate speed to produce a tangential speed of between 30 and 180 m/min. The electric discharge (spark) is generated in the inter-electrode gap between the rotating wheel and workpiece which is immersed in a dielectric fluid. Generally, kerosene oil, paraffin oil, transformer oil or de-ionized water are used as dielectric fluid. A constant inter-electrode gap in the range of 10–75 µm is maintained by the servo system. A DC pulse generator is used for supplying DC electrical energy capable of providing voltage, current and discharge frequencies in the ranges of 30–400 V, 30–100 A and 2–500 kHz, respectively. Different machining configurations are possible based on the alignment of the rotational axis of the grinding wheel with the workpiece. Typically, three different configurations are possible, i.e. electrodischarge cutoff grinding, electrodischarge face grinding and electrodischarge surface grinding [1, 2].

Fig. 3.1 Working principle of EDG

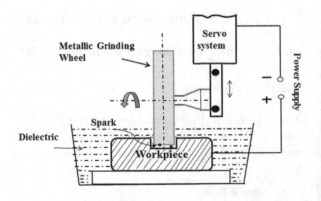

3.1.1.3 Process Mechanism and Parameters

During EDG, the anodic workpiece and a portion of the cathodic grinding wheel are immersed in the dielectric while being electrically in contact with the DC power supply. During pulse-on time and under the influence of the breakdown voltage, the dielectric breaks down and a conductive channel is formed in the inter-electrode gap between cathode and anode where rapid ionization takes place causing the first spark to occur at the location and where the inter-electrode gap is least. This leads to a sudden and nearly instantaneous localized temperature increase to between 8000 and 12,000 °C resulting in material removal from the electrodes by melting and vaporization. During the pulse-off time, breakdown of the spark occurs and fresh dielectric enters into the inter-electrode gap. During this period, the rotary motion of the grinding wheel ensures the effective ejection of the molten material. Due to the improved flushing efficiency, no debris accumulation takes place in the inter-electrode gap, resulting in an improved material removal rate and better surface finish as compared to conventional EDM and grinding processes.

The important process parameters that may influence EDG are as follows: discharge voltage, peak current, pulse-on time, pulse-off time (or pulse frequency and duty factor), polarity, type of dielectric and rotational speed of grinding wheel. The use of a positive electrode polarity, higher peak currents and longer pulse-on times leads to a higher material removal rate, whereas a negative electrode polarity, lower peak currents and shorter pulse-off times are preferred for improved surface finish.

3.1.1.4 Applications

EDG is generally used for machining of polycrystalline diamond (PCD), ceramics and metallic composites. This process is capable of achieving surface roughness value up to 0.2 μm and dimensional tolerance up to ±2.5 μm. It can machine extremely hard materials 2 to 3 times faster than conventional grinding [3].

3.1.2 Electric Discharge Abrasive or Diamond Grinding (EDAG or EDDG)

3.1.2.1 Process Details

Electric discharge abrasive grinding (EDAG) is an important variant of EDG where a metal-bonded grinding wheel with embedded abrasive particles is used to enhance the machining productivity while also achieving an improved surface finish especially when machining polycrystalline diamond (PCD), ceramics, sintered carbides and metallic composites. In this process, the material is removed by the interaction effect of the electric discharge erosion of EDM and the mechanical abrasion of the grinding process. Al_2O_3, SiC and CBN are commonly used abrasive in this process. When diamond is used as abrasive, this process is known as electric discharge diamond grinding (EDDG). The increased hardness and wear resistance of diamond are advantageous to the process and lead to enhanced machining characteristics [3, 4].

Figure 3.2a shows the working principle and mechanism of material removal in the EDAG process. The metal-bonded abrasive/diamond grit grinding wheel is electrically connected to the negative terminal (therefore acts as cathode), whereas the workpiece is connected to the positive terminal (therefore acts as anode) of DC pulse power supply. A small portion of the workpiece and the abrasive grinding wheel are immersed in the dielectric, and a constant inter-electrode gap is maintained between them by a servo system. During the pulse-on time, the spark occurs in the IEG between anodic workpiece and cathodic abrasive wheel under the influence of discharge voltage softening the work material and removing some material by melting and vaporization. During the pulse-off period, the molten material is flushed from the inter-electrode gap under the rotary motion of grinding wheel, while mechanical abrasion occurs due to the impregnated abrasive particles on the wheel surface and by dielectric flushing.

Figure 3.2b illustrates the sequence of spark erosion and abrasion for removal of workpiece material. After the collapse of the spark, the abrasive particles come in contact with the workpiece, thereby performing micro-cutting and abrasion of the thermally softened workpiece material along with the recast layer (if any) from the workpiece surface. This cycle repeats itself until the required geometry is attained. The introduction of abrasive particles (especially hard and wear resistant diamond particles) enhances the material removal rate and improves the finish as compared to EDG and conventional grinding. It is also worth mentioning that self-dressing of

Fig. 3.2 **a** Working principle and **b** mechanism of material removal in EDAG

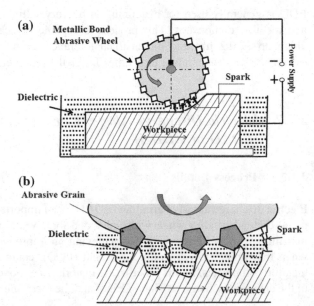

the grinding wheel takes place during the electric discharge phase of EDDG and therefore increases its efficiency. It may therefore be specifically used for precise truing/dressing of metal-bonded abrasive grinding wheels [3].

In addition to those significant process control parameters as identified for EDG, type and size of abrasive particles, the bond material may also influence the EDAG/EDDG process.

Extensive research has been done on the effect of the process control parameter on the productivity and product quality of EDAG and EDDG [4–8]. Rajurkar et al. [4] used a copper-bonded diamond wheel to machine Al-SiC composite and titanium alloy. They found that the material removal rate may be up to five times greater than EDM and up to twice that of conventional grinding. Kozak [5] demonstrated an improved surface finish with low current, high wheel speed and finer abrasive particle grit size. Aoyama and Inasaki [6] reported a decrease in grinding forces and an increase in wheel wear with higher voltage settings. Koshy et al. [7] investigated the mechanism of EDDG and put forward the suggestion that the reduction of the grinding forces and specific cutting energy is the result of thermal softening of the workpiece caused by the electric discharge. Singh et al. [8] observed a reduction in material removal rate at low current and high speed due to glazing of the diamond abrasive impregnated grinding wheel while machining WC-Co composite and HSS workpieces by EDDG.

3.1.2.2 Applications

The main application of EDAG/EDDG is to machine certain advanced and difficult-to-machine materials such as PCD, ceramics, carbides, metal matrix composites and titanium alloys with high material removal rates and good surface integrity. This process is also used for dressing/truing of metal-bonded abrasive/diamond wheels. Machining rates of up to 270 mm^3/min and an average surface roughness up to 0.2 μm can be achieved with this process.

3.2 EDM Combined with ECM

3.2.1 Electrochemical Discharge Machining (ECDM)

3.2.1.1 Introduction

The electrochemical discharge machining process (ECDM) is a combination of electric discharge machining (EDM) and electrochemical machining (ECM). This process is also known as electroerosion dissolution machining (EEDM), electrochemical spark machining (ECSM), electrochemical arc machining (ECAM) and spark-assisted chemical engraving (SACE). ECDM has found broad appeal as a micro-machining process that is able to overcome the difficulties of micro-machining of glass and has also further found a wide variety of applications including machining and drilling of micro-holes in non-conducting materials such as ceramics.

The process involves a complex combination of electrolytic dissolution and spark erosion which causes a breakdown of the insulating layer of gas bubbles formed during electrochemical reactions in the vicinity of the tool tip after application of an appropriate DC voltage between the cathodic tool and an auxiliary anodic electrode [9]. The combination of EDM and ECM results in smoother surfaces with minimum recast layer, higher material removal rate and lower tool wear.

3.2.1.2 Equipment and Working Principle

The schematic of a typical ECDM setup is shown in Fig. 3.3. It consists of a workpiece immersed in a tank of electrolyte, tool cathode, an auxiliary electrode as anode and a DC power supply capable of supplying up to 120 V. An aqueous solution of sodium hydroxide or potassium hydroxide is commonly used as electrolyte (other electrolytes are sodium chloride, sodium nitrate and hydrochloric acid). Stainless steel, nickel and copper are preferred as electrode materials because of their excellent resistance against corrosion. The cathodic tool is immersed in the electrolyte solution, and the level of the electrolyte is kept just 2 to 3 mm above the tool tip. The auxiliary electrode (anode) is also immersed in the electrolyte but kept at a higher level commensurate with a minimum tool electrolytic dissolution. Electrolysis commences when a suitable DC voltage is applied between the cathodic tool and auxiliary electrode. An electric discharge (spark) will occur upon the application of a potential difference (voltage) in excess of the critical value along with an adequate induced critical current density at the tool. This results in material removal from the workpiece due to melting and vaporization caused by the heat energy of the spark discharge [9–12].

A typical voltage and current profile during the ECDM process is shown in Fig. 3.4. As the voltage between the cathode and anode is increased, hydrogen

Fig. 3.3 Schematic of ECDM process

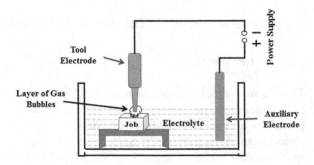

Fig. 3.4 Typical voltage and current profile during ECDM [9], with kind permission from Elsevier

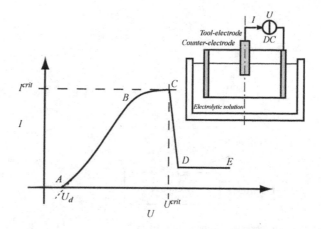

bubbles evolve at the cathode (region A–C). At a critical voltage and current (point C), electric discharge commences after the gas bubble insulating layer is broken down. Beyond point C, the current drops and discharging continues at the tool surface inside the electrolyte. The current and voltage corresponding to the critical point are known as the critical voltage and critical current.

When a travelling wire is used as the tool electrode, i.e. cathode, the process is known as wire electrochemical discharge machining (WECDM) or travelling wire electrochemical spark machining (TW-ECSM) [13, 14].

3.2.1.3 Process Mechanism and Parameters

The process mechanism of ECDM involves sequential removal of material by ECM and EDM. When a suitable DC voltage is applied between the cathodic tool electrode and the auxiliary anodic electrode, both immersed in an electrolyte, electrolysis reactions cause reduction of the electrolyte liberating hydrogen at the cathode. When the applied voltage is increased beyond a threshold value, then more hydrogen gas and vapour bubbles evolve and grow in size. Density of gas bubbles becomes so high that they coalesce into a gas film, which isolates the tool electrode from the electrolyte. In this film, the electric field is high enough to allow spark discharges to occur at the tip of cathodic electrode and the electrolyte due to the breakdown of the insulating layer of gas bubbles under the influence of the applied DC voltage. The heat generated by the spark discharges, and probably, some chemical etching contribute to the erosion and removal of material from the non-conducting workpiece. In other words, atoms are dislodged from the workpiece material due to the intense heating caused by the bombardment of electrons on the workpiece surface [9–11]. At the same instant, the contact between the tool and the electrolyte re-establishes due to the disturbance in the bubble geometry and electrolysis reactions takes over again, bubble gets built up, and the cycle keeps

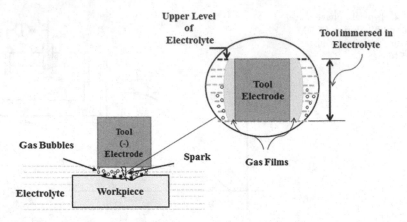

Fig. 3.5 Process mechanism of ECDM

repeating itself [12]. There may be chances of occurrence of some partial sparks based on gas film characteristics, which do not have enough energy to cause the material removal. These hamper the efficiency of the process and also affect the surface finish.

Gas bubble coalescence and the spark discharging phenomena in the ECDM process are shown in Fig. 3.5. Due to heating of the electrolyte, the generation of hydrogen gas bubbles and formation of gas film take place around the cathode. The characteristics of the hydrogen gas film play a key role in machining during the ECDM process. The immersed tool electrode is insulated by the gas film due to bubble coalescence in the electrolyte [10]. Due to the insulation, sparks are generated only from the tip of the tool. A stable and dense gas film suits the discharging process, whereas an unstable film results in fluctuating sparking and non-repeatable machining [9].

The basic principle of WECDM is the same as that of ECDM (i.e. softening, melting and etching), but continuous flushing of the electrolyte is essential to avoid the wire breakage. Mixing of SiC powder in the electrolyte (mostly KOH) improves surface quality of the workpiece by polishing and finishing [15].

Supply voltage, electrolyte type and concentration, inter-electrode gap, shape, size and material of the tool are the most important process parameters of the ECDM process. In addition, wire material and speed are important parameters for WECDM. Electrolyte type, applied voltage and SiC particle size (only in WECDM) are the most influencing factors for surface roughness of the electrochemical discharge machined workpiece.

3.2.1.4 Applications

ECDM as a micro-machining process has a wide variety of applications for machining various micro-features (i.e. micro-holes, micro-cavities, 3D micro-shapes) especially in non-conductive materials such as various types of composites and ceramics (aluminium oxides, zirconium oxides, silicon nitride, glass, quartz, pyrex, etc.) which are frequently used in micro-electromechanical systems (MEMS) and other micro-devices.

ECDM of non-conducting ceramics was carried out by Doloi et al. [16] with significant improvement in material removal rate and surface quality. Jain et al. [17] reported the use of a novel process named as electrochemical spark abrasive drilling (ECSAD) for machining alumina and borosilicate glass to enhance the capability of the ECDM process by using an abrasive cutting tool. The abrasive particles greatly assisted the ejection of the molten material from the machining zone. Significant improvements in material removal rate and machining depth were observed for both alumina and glass when compared to the conventional ECDM process. Moreover, the ECSAD process performance was observed to significantly improve as a function of an increased supply voltage. Chak and Rao [18] investigated the use of a cylindrical abrasive electrode (containing diamond particles as abrasive), during a trepanning operation of Al_2O_3 subjected to a pulsed DC voltage, which resulted in reduction of the tendency of micro-crack formation on machined surface at greater machined depth and enhanced the cutting ability and work surface quality. Jui et al. [19] successfully machined a micro-hole with an aspect ratio of 11 in a 1.2-mm-thick glass plate with 250 nm average surface roughness using an in-house developed tungsten micro-tool. Complex 3D micro-structures (features less than 100 μm) have also been created in Pyrex glass [20, 21] using ECDM. Slitting and slicing of glass and quartz are major applications of the WECDM process [22, 23].

3.2.2 Electrochemical Discharge Grinding (ECDG)

3.2.2.1 Process Details

Electrochemical discharge grinding (ECDG) hybridizes spark erosion, electrolytic dissolution and mechanical abrasion to machine electrically conductive workpieces. This hybridization results in a defect free and highly finished work surface [24]. Figure 3.6 illustrates the basic working principle and material removal mechanism of ECDG.

A metal-bonded abrasive grinding wheel is electrically connected as the cathode (negative terminal of a pulsed power supply), while the workpiece acts as the anode (connected to the positive terminal). The electrolyte flows into the inter-electrode gap between the rotating wheel and the workpiece, and a constant gap is maintained between them by a servo system. Initially, dissolution occurs at the workpiece (anode) due to the electrochemical reaction induced by the pulsed power supply. The dissolved material is flushed away by the electrolyte. A passivating oxide layer is also formed on the work surface during the electrochemical reaction. In the

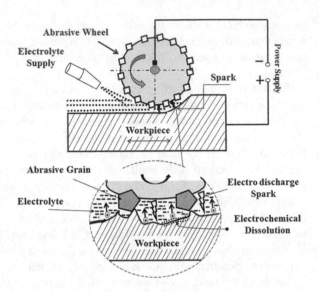

Fig. 3.6 Working principle and mechanism of ECDG

subsequent electroerosion phase, the spark discharges depassivate the oxide layer, therefore enhancing the effectiveness of the electrochemical dissolution and thereby enhancing the workpiece material removal rate. The spark discharges also cause thermal softening, melting and vaporization of workpiece material that is subsequently removed from the machining zone by the continuous flushing of the electrolyte and by the mechanical action of the grinding wheel abrasives. The mechanical abrasion ensures removal of any recast layer formed during the discharge erosion phase and completely depassivates the oxide layer (if present). The electrochemical dissolution phase smoothes the cratered surface resulting from the electrodischarge erosion and therefore improves the surface finish. The spark discharge also performs truing/dressing of grinding wheel by removal of the glazed layer during machining and helps to maintain the accuracy of the process. Sodium and potassium nitrates are the most commonly used electrolyte solutions in ECDG. Instead of abrasive wheel, a simple graphite grinding wheel may also be used in certain ECDG applications.

3.2.2.2 Applications

ECDG is used for grinding of carbide cutters, honeycomb structures, metal matrix composites and other fragile parts. This process is also used for truing/dressing of grinding wheels and micro-grinding of tools [24–26]. Surface roughness values of between 0.1 and 0.8 µm, and production tolerances of ±30 µm are attainable in ECDG.

References

1. Abothula BC, Yadav V, Singh GK (2010) Development and experimental study of electro-discharge face grinding. Mater Manuf Process 6:482–487
2. Yadav RN, Yadav V (2012) Electrical discharge grinding (EDG): a review. Paper presented at the national conference on trends and advances in mechanical engineering, YMCA University of Science & Technology, Faridabad, 19–20 October 2012
3. Jain VK (2002) Advanced machining processes. Allied Publishers Pvt. Ltd., New Delhi
4. Rajurkar KP, Wei B, Kozak J, Nooka SR (1995) Abrasive electro-discharge grinding of advanced materials. In: Proceeding of the 11th international symposium of electro-machining (ISEM-11), Lausanne, pp 863–870
5. Kozak J (2002) Abrasive electro discharge grinding (AEDG) of advanced materials. Arch Civil Mech Eng 2:83–101
6. Aoyama T, Inasaki I (1986) Hybrid machining-combination of electrical discharge machining and grinding. In: Proceeding of the 14th North American manufacturing and research conference, SME, pp 654–661
7. Koshy P, Jain VK, Lal GK (1996) Mechanism of material in electrical discharge diamond grinding. Int J Mach Tool Manuf 36(10):1173–1185
8. Singh GK, Yadava V, Kumar R (2010) Diamond face grinding of WC-Co composite with spark assistance: experimental study and parameter optimization. Int J Precis Eng Man 11:509–518
9. Wuthrich R, Fascio V (2005) Machining of non-conducting materials using electrochemical discharge phenomena—an overview. Int J Mach Tool Manuf 45:1095–1108
10. Bhattacharyya B, Doloi BN, Sorkhel SK (1999) Experimental investigations into electro chemical discharge machining (ECDM) of non-conductive ceramic materials. J Mater Process Techol 95:145–154
11. Wuthrich R, Hof LA (2006) The gas film in spark assisted chemical engraving (SACE)-a key element for micromachining applications. Int J Mach Tool Manuf 46:828–835
12. Kulkarni AV (2012) Electrochemical spark micromachining processes. In: Kahrizi M (ed) Micromachining techniques for fabrication of micro and nano structures, 1st edn. InTech, Shanghai, pp 235–251
13. Tsuchiya H, Inoue T, Miyazaiki M (1985) Wire electrochemical discharge machining of glass and ceramics. Bull Jpn Soc Precis Eng 19(1):73–74
14. Wuthrich R, Ziki JDA (2015) Micromachining using electrochemical discharge phenomenon. Elsevier, Oxford
15. Kuo KY, W KL, Yang CK, Yan BH (2015) Effect of adding SiC powder on surface quality of quartz glass micro-slit machined by WECDM. Int J Adv Manuf Technol 78:73–83
16. Doloi B, Bhattacharyya B, Sorkhel SK (1999) Electrochemical discharge machining of non-conducting ceramics. Def Sci J 49(9):331–338
17. Jain VK, Choudhury SK, Ramesh KM (2002) On the machining of alumina and glass. Int J Mach Tool Manuf 42:1269–1276
18. Chak SK, Rao V (2007) Trepanning of Al_2O_3 by electro-chemical discharge machining (ECDM) process using abrasive electrode with pulsed DC supply. Int J Mach Tool Manuf 47:2061–2070
19. Jui SK, Kamaraj AB, Sundaram MM (2013) High aspect ratio micromachining of glass by electrochemical discharge machining (ECDM). J Manuf Proc 15:460–466
20. Zheng ZP, Cheng WH, Huang FY, Yan BH (2007) 3D micro-structuring of pyrex glass using the electrochemical discharge machining process. J Micromech Microeng 17:960
21. Cao XD, Kim BH, Chu CN (2009) Micro-structuring of glass with features less than 100 μm by electrochemical discharge machining. Precis Eng 33:459–465
22. Peng WY, Liao YS (2004) Study of electrochemical discharge machining technology for slicing on-conductive brittle materials. J Mater Process Techol 149:363–369

23. Kuo KY, W KL, Yang CK, Yan BH (2013) Wire electro chemical discharge machining (WECDM) of quartz glass with titrated electrolyte flow. Int J Mach Tool Manuf 72:50–57
24. Benedict GF (1987) Nontraditional manufacturing processes. Marcel Dekker Inc., New York
25. Wei C, Hu D, Xu K, Ni J (2011) Electrochemical discharge dressing of metal bond micro-grinding tools. Int J Mach Tool Manuf 51(2):165–168
26. Liu JW, Yue TM, Guo ZN (2013) Grinding aided electrochemical discharge machining of particulate reinforced metal matrix composites. Int J Adv Manuf Technol 68:2349–2357

Chapter 4
Assisted Hybrid Machining Processes

Abstract The growing trend of miniaturization of products and rapid development in precision engineering demand higher machining quality of micro-holes, micro-grooves and complicated 3D micro-structures in parts of especially difficult-to-machine materials. Assisted-type HMPs were developed specifically to overcome the limitations of advanced machining processes and to improve the machining quality and productivity. Assisted-type HMPs are another important category of HMPs whereby assistance of an external source is used to overcome the limitations of the primary material removal process to facilitate efficient and effective machining. This chapter discusses working principles, process mechanisms and typical applications of some important vibration, heat, abrasive and magnetic field-assisted hybrid machining processes.

Keywords Vibration · Laser · Abrasive · Assisted HMP · Flushing efficiency · Electrochemical dissolution · Micro-machining

Assisted HMPs are the combinations of two or more processes, where an advanced machining process is introduced to assist the primary material removal process to enhance the machinability. The primary process may be of electrochemical, electrodischarge or conventional mechanical type. Generally, high-frequency vibrations, heat (such as lasers), abrasives (or a fluid medium) and magnetic fields are used for assistance. These are then referred to as vibration-assisted machining, laser-assisted machining, abrasive-assisted machining and magnetic field-assisted machining, respectively. Precise machining of ceramics, composites and difficult-to-machine metals and alloys, fabrication of micro-parts, shaping high aspect ratio cavities, etc. are some of the specialized applications of assisted HMPs.

© The Author(s) 2016
K. Gupta et al., *Hybrid Machining Processes*,
Manufacturing and Surface Engineering,
DOI 10.1007/978-3-319-25922-2_4

4.1 Vibration-Assisted HMPs

Vibration-assisted HMPs are an important type of assisted HMPs where high-frequency and low-amplitude vibrations are used to assist the primary material removal process. The concept of using high-frequency vibrations for material removal purposes has been known for more than half a century [1, 2] and generally referred to as ultrasonic machining (USM). During USM, an assembly of a transducer and concentrator is used to propel abrasive particles suspended in slurry to erode the workpiece surface. Ultrasonic-assisted machining, on the other hand, utilizes directly transmitted ultrasonic vibrations (frequency range: 10–40 kHz; amplitude range: 1–200 μm) obtained with a piezoelectric or piezomagnetic transducer to the tool or workpiece of the primary machining process such as electrodischarge machining (EDM), drilling, turning, milling or grinding in the absence of an abrasive slurry to enhance the process. However, in ultrasonic-assisted electrochemical machining (USECM), abrasives are used along with the electrolyte and are largely responsible for achieving the required material removal rate. In most cases, ultrasonic-assisted machining relies on superposition of high-frequency-induced vibrations (preferably along the main cutting direction) on the primary tool motion to enhance the machining process. Typically, vibrations with frequencies of up to 25 kHz and amplitude of up to 15 μm are employed. The advantages of ultrasonic-assisted machining are avoids chipping or destruction of the surface layer of the tool, decreases the cutting force, improves the tool life, improves the surface quality of the product and improves process productivity. Figure 4.1 shows a schematic of a typical USM system which consists of a

Fig. 4.1 Schematic diagram of a typical USM system

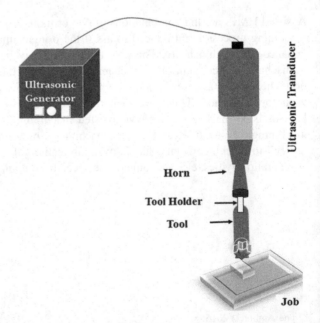

high-frequency signal generation system, a transducer to convert the signal, high-frequency mechanical vibrations and an acoustic horn to amplify the ultrasonic vibrations to then transfer them to the attached tool thereby facilitating enhanced cutting of the job. Two types of ultrasonic vibration-assisted machining exist:

1. **Ultrasonic-assisted mechanical machining**: in this technique, ultrasonic vibrations are used to enhance conventional machining techniques including turning, drilling, grinding and milling operations;
2. **Ultrasonic-assisted advanced machining**: where performance of advanced machining processes such as ECM and EDM is enhanced by using ultrasonic vibrations.

4.1.1 Ultrasonic-Assisted ECM (USECM)

4.1.1.1 Introduction

Electrochemical machining is mostly ineffective with regard to the machining of difficult-to-cut materials such as composites; because of its inherent process requirement, the workpiece material needs to be electrically conductive. Composite materials may be engineered with the enhanced properties in selected preference directions and may contain both metallic and non-metallic constituents. Challenges of quality and productive machining of such materials can be met by hybridizing two advanced processes, one for efficient removal of metallic conducting constituent such as ECM process and other process for efficiently removal of non-conductive constituent (which may be hard and/or brittle) such as USM. This specific combined hybrid machining process is referred to as ultrasonic-assisted electrochemical machining (USECM). It offers many advantages over its constituent processes (ECM and USM) when used in isolation, such as improved work surface quality, lower tool wear and higher productivity.

4.1.1.2 Equipment and Working Principle

A typical USECM system along with its various components is shown in Fig. 4.2. It includes an ECM system consisting of a DC power supply, tool electrode and its feed mechanism; components of ultrasonic system viz. ultrasonic transducer (preferably piezoelectric) coupled with ultrasonic generator, horn with a tool holder to transmit the ultrasonic energy to the tool and for amplifying the tool stroke, and an electrolyte containing an appropriate abrasives.

Usually, a DC voltage of between 2 and 30 V is applied across the anodic workpiece and cathodic tool for material removal by electrochemical action in this process. Ultrasonic vibrations, with frequency and amplitude ranges of 10–20 kHz and 10–40 μm, respectively, are induced in the tool. Electrochemical dissolution

Fig. 4.2 Schematic
representation of a typical
USECM set-up

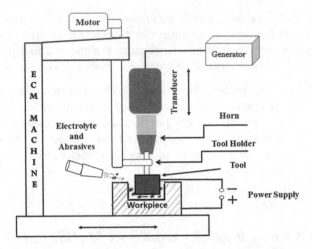

and the formation of a passivating oxide layer occur on the workpiece surface by
ion formation and movement within the electrolyte producing a high intensity
current flow. The abrasive particles are accelerated by the vibro-impact motion of
the tool electrode that is driven by the ultrasonic transducer. The passivating layer is
then removed by these ultrasonically accelerated abrasives that impact the workpiece surface. This process also helps to maintain a constant inter-electrode
gap. The cycle continues until the required work geometry or shape is formed.

4.1.1.3   Process Mechanism and Parameters

The success of ECM depends on material removal from the anode (workpiece)
surface by electrochemical dissolution. This anodic dissolution phase is accompanied by the formation of a passivating metal oxide layer on the workpiece surface.
Removal of this layer is mandatory because it hinders further electrolytic dissolution. To remove this oxide layer and other compounds from the anode surface,
assistance is provided by the ultrasonically vibrated particles thereby improving
productivity of the ECM process. These abrasive grains depassivate the machined
surface of the workpiece mechanically by removing the brittle oxide layer and
increase the rate of the electrochemical dissolution. The ultrasonic vibrations
intensify the ECM action by increasing the diffusion of metal ions and influence the
ECM process by the following:

(i) Facilitating the removal of heat and reactions products from the machining
     area;
(ii) Controlling the rate of the passivation process by direct impingement of
     abrasive particles on the electrode and machined surface; and consequently
     increasing the machining efficiency by ensuring higher machining speeds and
     lower tool wear compared to conventional ECM and USM.

The main process parameters that affect material removal rate, accuracy and surface finish of USECM are amplitude and frequency of the ultrasonic vibrations; abrasive type and size; electrolyte type, concentration, temperature and flow rate; electrode feed rate; and voltage; current and inter-electrode gap.

Aluminium oxide, boron carbide and silicon carbide are the most extensively used abrasives. Sodium chloride, sodium nitrate and potassium nitrate are the commonly used electrolytes in USECM.

4.1.1.4 Applications

USECM is basically used for machining difficult-to-machine materials especially ceramics and to shape free-form surfaces and fabrication of micro-parts. This process aims to improve the performance of ECM and thereby increasing the rate of electrochemical dissolution which implies an increase in the material removal rate by improving the hydrodynamic effects and effective removal of heat to also improve work surface quality, etc. [3, 4]. Skoczypiec and Ruszaj [5] improved both the productivity and the surface roughness of ECM of titanium alloys by introducing 4 μm ultrasonic vibrations to the tool electrode in a sodium nitrate electrolyte. Pa [6] used the USECM process to improve the surface finish of large holes. Recently, Ghoshal and Bhattacharyya [7] fabricated tungsten micro-tools using ultrasonic-assisted micro-ECM and reported an increase in volumetric material removal rate with the increase in vibrational amplitude.

4.1.2 Ultrasonic-Assisted EDM (USEDM)

4.1.2.1 Introduction

The hybridisation of EDM and USM is known as ultrasonic-assisted electric discharge machining (USEDM). Certain inherent limitations of the EDM process such as ineffective machining debris flushing, formation of recast layers and frequent contact between tool and workpiece may be overcome by assistance of an additional process [8]. The combination of EDM and USM results in considerable benefits for machining of hard and brittle materials such as increased flushing efficiency, lower tool wear rate, increased material removal rate and improved surface finish. In other words, employing ultrasonic vibrations is an effective way to scavenge debris from the machining zone, improve flushing efficiency and consequently improve material removal rate and geometric accuracy. Micro-versions of this process are extensively used for various industrial and scientific applications [9, 10].

4.1.2.2 Equipment and Working Principle

The ultrasonic vibrations inherent to the USEDM can be applied to any of three main elements of the EDM process, i.e. tool electrode, workpiece or dielectric fluid. Figure 4.3 depicts a schematic of a typical USEDM system whereby ultrasonic vibrations are applied to the tool electrode. It consists of an ultrasonic transducer coupled with a conventional EDM set-up, which has a tool electrode (acting as cathode) attached to a spindle rotated by a motor, workpiece (acting as anode) submerged in a dielectric tank and a power supply unit. An ultrasonic transducer is connected to the spindle through a connector and provides mechanical vibrations to the tool that is superimposed on the feed movement.

A typical EDM process commences between the tool electrode and the workpiece at appropriate settings of its process parameters (i.e. voltage, current, pulse-on and pulse-off time). High-frequency mechanical vibrations in a range of 20–60 kHz and with an amplitude of 2–10 μm are employed either to the tool (as shown in Fig. 4.3) or to the workpiece. In certain cases, lower frequency vibrations are also used. The conventional EDM process produces the material removal from the workpiece leaving tiny craters on its surface while the ultrasonic vibrations assist to flush the machining debris efficiently from the inter-electrode gap thereby facilitating further discharge and therefore achieving a higher material removal rate and stable process conditions.

Fig. 4.3 Schematic diagram of a typical USEDM system

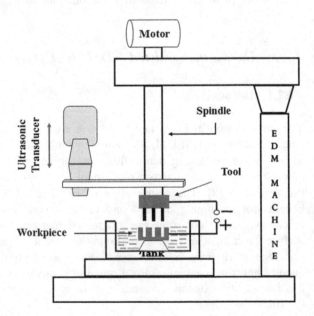

4.1.2.3 Process Mechanism and Parameters

The mechanism of USEDM integrates the mechanisms of EDM and USM. The main aim of this integration is enhanced scavenging (which is a significant problem in conventional EDM) of the machining debris from the inter-electrode gap between the tool electrode and the workpiece with the help of ultrasonic vibrations to improve productivity and part quality. Melting and vaporization of the workpiece material by the plasma channel takes place during the EDM phase of the USEDM process which is accompanied by the simultaneous flushing of the debris by a tool subject to mechanical vibration. Due to high-frequency vibrations, the tool momentarily displaces towards the workpiece surface with a subsequent increase in the electric field intensity.

At a sufficiently small inter-electrode gap, the dielectric breaks down, a plasma channel is formed and particles are ejected from the workpiece surface leaving tiny craters (Fig. 4.4). During pulse-off, the tool displaces away from the surface whereby increasing the inter-electrode gap, the current drops and the plasma channel collapses. The periodic relative movement between the tool and the workpiece due to the ultrasonic vibration causes a dielectric flow and subsequent effective flushing of the adhered debris particles, which prevents further settlement of other debris on the workpiece surface.

The improved flushing conditions and homogenization of the dielectric consequently result in faster and more stable process with improved accuracy. The cycle of relative motion (up and down) of the tool continues until the required shape or geometry is attained.

The process stability, productivity and part quality are directly influenced by the ultrasonic vibration amplitude and frequency, tool material and dielectric type.

4.1.2.4 Applications

Important applications of USEDM are: machining of hard and brittle materials such as carbides and ceramics, high aspect ratio holes and deep cavities, fabrication of micro-parts and electrodes for EDM itself. Abdullah et al. [11, 12] reported

Fig. 4.4 Mechanism of material removal in the USEDM process

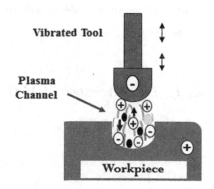

improved surface integrity aspects and higher material removal rates as compared to conventional EDM. Zhao et al. [13] reported improvement in machining efficiency, stability and workpiece surface quality when using USEDM to drill micro-holes of diameters less than 0.2 mm and with aspect ratios of more than 15 in titanium alloys. Uhlmann and Domingos [14] reported a reduction of 21 % in tool wear rate and 11 % increase in material removal rate when machining of deep cavities of aspect ratio 12 in high temperature resistant material MAR-247 (as used for turbine components) when using USEDM.

4.2   Heat-Assisted HMPs

The adverse effects on machinability of certain mechanical properties of the material to be machined such as high hardness, strength, toughness and stiffness may be largely overcome by an increase in localized workpiece temperature. Heat-assisted HMPs rely on this fact to improve the machinability of especially difficult-to-machine metals and alloys by machining them at elevated temperatures by utilizing an external heat source. The use of an external heat source improves the machinability by minimizing the machining forces, improving the work surface integrity and enhancing the tool life. This heat source may be in the form of a laser beam, electron beam, plasma beam, high-frequency induction or electric current etc.

Laser-assisted machining is one of the important and most widely used category of the heat-assisted HMPs, in which a laser beam directed (i) to soften the workpiece during conventional machining, i.e. turning and milling; (ii) to assist the electrochemical dissolution in ECM; and (iii) to machine the workpiece roughly to the required shape and geometry prior to their final machining by EDM. Different types of lasers such as ruby, CO_2 and Nd-YAG are used to meet different requirements of work material and process. Laser-assisted HMPs are of two types:

1. **Laser-assisted mechanical machining**: in which, a laser is used to heat the workpiece ahead of the cutting tool during conventional machining processes such as turning, milling and grinding; and
2. **Laser-assisted advanced machining**: where laser assistance enhances the material removal in electrolytic dissolution and electrodischarge-based AMPs.

The following subsections details laser-assisted ECM and laser-assisted EDM.

4.2.1   Laser-Assisted ECM (LAECM)

4.2.1.1   Introduction

The use of a laser is one of several ways to improve the precision and efficiency of the ECM process by effectively controlling the electrochemical dissolution. The primary role of a laser in ECM is to improve the localization of the dissolution process. The main mechanism of material removal in laser-assisted ECM (LAECM) is enhanced by electrolytic dissolution because of an improved thermal activation brought upon by a focused laser beam. Additionally, it also helps in removing the passivating metallic oxide layers formed on the workpiece due to the evolution of oxygen at the anode during the electrolysis process. These factors noticeably improve material removal rate, geometric accuracy and surface quality of the parts. This process is also generally referred to as laser-assisted jet electrochemical machining (LAJECM).

4.2.1.2 Equipment and Working Principle

The schematic of a typical LAECM system is shown in Fig. 4.5. It consists of three main systems: (i) electrolyte supply system; (ii) power supply system; and (iii) laser assistance system.

The electrolyte and the focused laser beam are coaxially aligned and pass through a thin tube-shaped nozzle which acts as the cathodic tool and creates a jet of electrolyte. This combined coaxial laser beam and electrolyte jet impinge the anodic workpiece surface at the same location. The laser beam causes a localized increase of temperature in the workpiece and electrolyte. Upon a suitably applied voltage between the tool and the workpiece, the ECM process is initiated more readily and the diffusion of ions and the transportation of the removed workpiece material are enhanced. This improves process productivity and machining quality. Electrolyte velocity is varied by compressed air. The inter-electrode gap (distance between nozzle exit and work surface) varies between 0.1 and 2 mm. Generally, a

Fig. 4.5 A schematic of the LAECM process

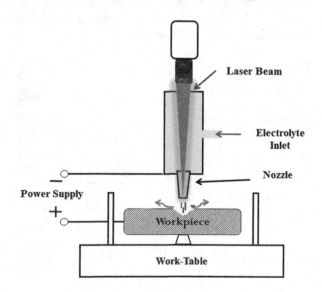

DC power supply unit with a voltage and current capacity of 0–200 V and 0–20 A is used. A low-power Nd-YAG laser is most frequently used. Pulsed lasers are extensively used for micro-machining applications by LAECM.

4.2.1.3 Process Mechanism and Parameters

LAECM combines two different sources of energy simultaneously i.e. electrical energy (ion formation) and light (photons). The primary mechanism of material removal is electrolytic dissolution, which is supported by the simultaneous action of a low-power laser beam that enhances electrolytic dissolution by its thermal activation. Heating by laser beam causes numerous physical and chemical phenomena to the machined surface as well as to the surface layer of the workpiece material. It enhances the kinetics of the electrochemical reactions providing faster dissolution. It also aids in breaking down the passivating metallic oxide layer found on some materials in certain electrolytes that inhibit further efficient dissolution.

Figure 4.6 illustrates the mechanism of LAECM. A focused laser beam in the machining area increases electrolyte temperature which in turn increases electrolyte conductivity and current density leading to an enhanced reaction in a localized area according to *Arrhenius law* [15, 16]. Electrochemical reactions are initiated more readily because higher temperatures lower the activation energy. The transportation process of reaction products by diffusion is also improved thereby decreasing the electrode polarization potential, which enables a higher current density to be achieved. Also, the local heating of the workpiece area subjected to the laser beam leads to changes in the equilibrium or open circuit potential. It increases the reaction rate that proceeds under the charge-transfer control, accelerates the mass transfer, changes the current efficiency, decreases the metal passivity degree and finally increases the current carrier concentration. The important process control parameters of LAECM are the inter-electrode gap, voltage, electrolyte concentration, electrolyte jet speed and laser power.

Fig. 4.6 Mechanism of material removal in the LAECM process [17], reproduced with kind permission from Elsevier

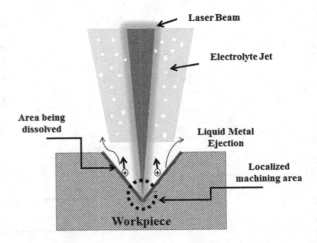

4.2.1.4   Applications

The main applications of LAECM include manufacturing of small holes, cavities and precise micro-shapes especially in the difficult-to-machine metals and alloys, composites and ceramics. Significant improvements in material removal rate (up to 54 %) were observed by Pajak et al. [16] using LJECM for Hastelloy, titanium alloy, stainless steel and aluminium alloy. Moreover, noticeable improvements in shape accuracy (up to 65 %) in terms of reduction in taper and surface finish with an average roughness up to 0.18 μm were also reported in holes and cavities produced in these materials. De Silva et al. [17] achieved average roughness up to 20 nm in 25-mm-deep cavities produced in titanium and Hastelloy when using LAECM. Due to electrolyte boiling, no detectable heat-affected zone or spark damage occurred. The laser also facilitated removal of brittle oxide layers enabling ECM of oxide forming metals (i.e. titanium) with benign electrolytes. Zang and Xu [18] successfully improved overall quality of laser drilled holes by removing recast layers and spatters by LAECM.

4.2.2   Laser-Assisted EDM (LAEDM)

4.2.2.1   Introduction

Laser-assisted electric discharge machining (LAEDM) is a hybrid process of laser beam machining (LBM) and EDM in which these processes are used sequentially, i.e. the laser beam is used to fabricate the required shape initially where after and EDM is used for final finishing. Lengthy machining times and high tool wear are the main drawbacks of EDM, whereas formation of a recast layer and heat-affected zones (HAZ) and low surface quality of the workpiece are the major limitations of LBM. Hybridization of LBM and EDM as LAEDM addresses their respective disadvantages. LAEDM is generally used in micro-machining applications for reducing production time and eliminating the recast layer and HAZ caused by laser ablation.

4.2.2.2   Working Principle and Applications

Figure 4.7 depicts the concept of LAEDM in which a short-pulsed laser beam (having pulse-on time ranging from nanoseconds to microseconds) capable of producing high ablation rates is used to roughly pre-machine the basic part features (i.e. groove, hole, cavity). This is subsequently followed by micro-EDM within a suitable dielectric to remove the surface defects caused by the thermal effects of laser ablation and to finish the feature in order to attain the required shape and quality. Nd-YAG lasers are most widely used in LAEDM because the main purpose of EDM in LAEDM is to perform the finishing operation only; therefore, tool wear

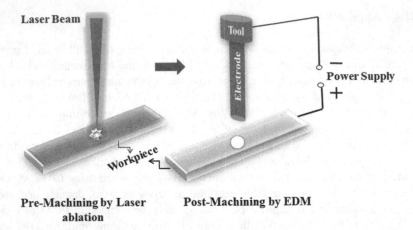

Laser Beam

Tool

Electrode

Power Supply
−
+

Workpiece

Pre-Machining by Laser   Post-Machining by EDM
ablation

Fig. 4.7 Concept of LAEDM process

is almost negligible and machining is faster. The efficiency, productivity and surface quality obtained with this process are much better than conventional LBM and miro-EDM when used in isolation.

EDM voltage; electrode type, material and feed rate; type of dielectric; laser power and pulse frequency are some of the important process parameters of LAEDM. Complicated 3D micro-structures [19], micro-grooves [20] and diesel fuel injection nozzle holes of 140 μm diameter [21] have been precisely machined by LAEDM.

4.3 Abrasive-Assisted HMPs

Various HMPs exist that involves assistance from some medium to enhance material removal rates and surface integrity, and reducing tool wear. These include abrasives, high-pressure lubrication coolants and magnetic field or magnetorheological (MR) fluids. This section introduces abrasive-assisted HMPs where the role of the abrasive particles includes the following:

(a) enhanced material removal by the primary process;
(b) improved scavenging of machining debris from the machining zone;
(c) scraping the recast layer from the work surface; and
(d) finishing the part by mechanical abrasion.

Silicon carbide (SiC), tungsten carbide (TiC) and aluminium oxide (Al_2O_3) are some important abrasives used in abrasive-assisted HMPs. Abrasive-assisted ECM (AECM) and abrasive-assisted EDM (AEDM) are the two most important abrasive-assisted HMPs and are therefore presented and discussed in the following sections.

4.3.1   *Abrasive-Assisted ECM (AECM)*

4.3.1.1   Process Details

AECM combines the mechanical abrasion action imparted by abrasive particles and electrolytic dissolution of ECM to manufacture largely defect-free and highly finished workpieces. Fundamentally, the abrasives polish the workpiece surface mechanically while it is being machined electrochemically.

Figure 4.8 illustrates the working principle of the AECM process. Non-conductive abrasive particles are suspended in the electrolyte that is flooded into the inter-electrode gap in between the electrically conductive cylindrical-shaped cathodic tool and anodic workpiece (in this case a ferrous material). Upon the application of a DC voltage across the inter-electrode gap, flooded with an appropriate electrolyte, current flows through the electrolyte from the workpiece to the electrode causing electrolytic dissolution of the workpiece material and formation of metal hydroxides as the product of the machining process. Rotary motion is imparted on the tool electrode while the workpiece is externally driven to reciprocate. This reciprocating motion of the workpiece also effectively transports the abrasives along the electrolyte.

During the finishing phase, a lump (or an accumulation) of abrasive particles are formed around the tool. This accumulation is formed due to the electrophoresis phenomenon that is caused due to the abrasive particles being positively charged and therefore being attracted to the negatively charged tool while suspended in a low conductivity electrolyte (see Fig. 4.8). Rotation of the accumulation of abrasives with the rotating electrode brings about effective removal of the metal hydroxides. The workpiece is finished by electrochemical dissolution and the mechanical action of the accelerating abrasive particles. Additionally, the abrasives also remove the undesirable passivating metal oxide layers therefore enhancing the efficiency of the process.

Fig. 4.8 Working principle of AECM process

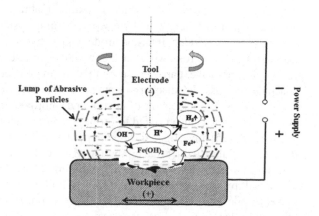

The process performance of AECM is a function of the following parameters: type, size, concentration and flow rate of the abrasive particles; type, concentration, temperature, pressure and flow rate of the electrolyte; rotary speed of the cathode; and reciprocating speed of the workpiece and voltage.

4.3.1.2   Applications

AECM is primarily used for machining small holes and micro-features with sharp edges and highly finished surfaces. Resizing and refinishing the pre-shaped/pre-machined holes are also important applications of AECM. Takahata et al. [22] used AECM to produce 3D micro-structures in stainless steel with a superfine surface finish (maximum roughness of 32 nm). Abrasive-assisted electrochemical grinding (AECG) is an important variant of AECM. Several researchers used AECG and reported on its beneficial outcomes. Zhu et al. [23] used AECG to produce precision holes of 0.6 mm diameter with sharp edges and without burrs. They also noted that AECG is an effective process to remove the recast layers. Kozak and Skrabalak [24] achieved significant improvements in material removal rate while machining super alloys, metal matrix composites and sintered carbides using metal-bonded grinding wheels with AECG.

4.3.2   Abrasive-Assisted EDM (AEDM)

4.3.2.1   Process Details

Abrasive-assisted EDM (AEDM) is a hybrid process in which abrasive particles are suspended in the dielectric that is circulated in the inter-electrode gap between tool and workpiece. Mechanical abrasion caused by the abrasive particles assists the EDM action and enhances further material removal. Figure 4.9 depicts the working principle of AEDM process in which a mixture of abrasive particles and dielectric is supplied to the machining zone by a separate pump system. The discharging occurs in the dielectric flooded inter-electrode gap between the cathodic tool and the anodic workpiece across which a DC potential difference is applied. During this process, the abrasive particles suspended in the dielectric reduce the electrical capacitance across the discharge gap by effectively increasing the gap size and ensuring improved dispersion of sparks and improved discharging characteristics. The tool electrode is also rotated to further increase the effectiveness of the process.

Lin et al. [25] investigated an important variant of AEDM by combining the advantages of abrasive jet machining (AJM) with EDM in a single process referred to as abrasive jet electrodischarge machining (AJEDM). Abrasive particles are transported by compressed air through a hollow tube-shaped electrode to the inter-electrode gap. Figure 4.10 illustrates the principles of AJEDM. The EDM action produces a series of electrical discharges that melts and vaporizes the

Fig. 4.9 Working principle of AEDM process

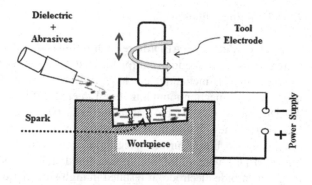

Fig. 4.10 Working principle of abrasive jet electrodischarge machining (AJEDM) process [25], reproduced with kind permission from Elsevier

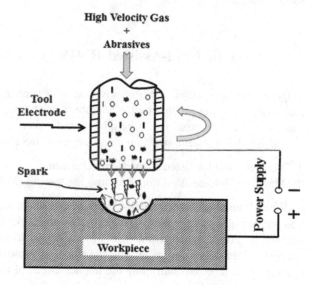

workpiece material. The abrasive particles transported by the compressed air eject the molten material and also remove the recast layer formed due to EDM. Removal of surplus material from the machining zone facilitates further discharging and increases the material removal rate along with an improved surface quality. It eliminates the need for subsequent finishing and trimming operations.

The important process parameters that affect AEDM are type, size, concentration and flow rate of the abrasive particles; pulse-on time; pulse-off time; type, pressure and flow rate of the dielectric; tool electrode and its feeding mechanism and power supply parameters.

4.3.2.2   Applications

Surfaces produced by AEDM are highly finished and free from recast layer. It is therefore used to produce mirror-finished complex shapes which are free from surface defects including cracks. AEDM is therefore extensively used for machining of injection moulding dies. Lin et al. [25] compared the machinability of AJEDM and dry EDM by machining SKD 61 steel (widely used for die manufacturing). They reported an improved material removal rate, lower surface roughness and better surface integrity aspects with AJEDM. Sun et al. [26] combined AEDM with ECM for drilling blind holes in stainless steel 304 using a tungsten carbide electrode. It resulted in high surface quality with a 15 nm average surface roughness.

4.4   Magnetic Field-Assisted HMPs

Magnetic field-assisted HMPs rely on the enhancement of the primary machining/finishing process by the addition (assistance) of magnetic field with the aim to improve workpiece surface quality and material removal rate.

This section presents two important magnetic field-assisted HMPs, i.e.

1. Magnetic field-assisted EDM (MAEDM); and
2. Magnetic field-assisted abrasive flow machining (MAAFM).

When combined with the EDM process, the magnetic field improves efficiency and surface quality by effective and rapid expulsion of machining debris from the machining zone. When combined with abrasive flow machining (AFM), the magnetic field assists by increasing the material removal rate and improving the surface finish through controlling the impact forces of the abrasive particles on the workpiece during finishing.

4.4.1   Magnetic Field-Assisted EDM (MAEDM)

Debris accumulation in the machining zone that adversely affects performance and efficiency has always been a perennial problem in EDM. The addition of a magnetic field to the process is one of several techniques that have been employed to address this challenge. Adding magnetic field enhances the process stability and increases the efficiency of the EDM component by effective removal of machining debris from the machining zone. This also improves the surface integrity of the EDMed workpiece and enhances the material removal rate [27].

The set-up of magnetic field-assisted EDM consists of a standard EDM machine coupled to a magnetic force-assisted device (Fig. 4.11). The magnetic force-assisted device comprises of a rotational disc fastened with two symmetric magnets and

Fig. 4.11 A typical set-up of magnetic field-assisted EDM process [27], reproduced with kind permission from Elsevier

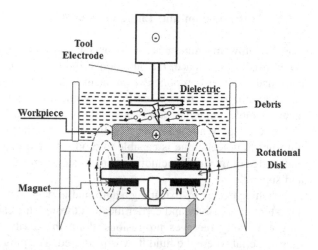

driven beneath the EDM process machining zone by electric motor. The introduction of the magnetic field exerts a magnetic force perpendicular to the electrode's motion. As a result, debris particle in the machining zone is subjected both to a magnetic force and to a centrifugal force. The resultant force which is the vector summation of the magnetic and centrifugal forces ensures effective and rapid flushing of debris particles from the inter-electrode gap and machining zone [27, 28].

The effective and rapid ejection of debris reduces the probability of abnormal discharge, and of debris to remelt and adhere on the machined surface thereby ensuring high workpiece surface integrity with minimum recast layer thickness. The presence of the magnetic force not only effectively ejects the debris, but also facilitates the settling of debris at the bottom of the fluid tank. The settling action accounts for the bulk of machining debris suspended in the machining gap.

Lin and Lee [27] observed an improvement of 300 % in material removal rate when machining SKD 61 by magnetic field-assisted EDM when directly compared to conventional EDM. Moreover, the presence of surface cracks and the recast layer thickness were both reduced. This process can produce higher aspect ratio holes compared with conventional EDM. Yeo et al. [29] reported on fabrication of high aspect ratio micro-holes in a hardened tool steel workpiece by magnetic field-assisted micro-EDM. They achieved a 26 % increase in machining depth as compared to conventional micro-EDM under similar working conditions.

4.4.2 Magnetic Field-Assisted Abrasive Flow Machining (MAAFM)

A brief overview of the abrasive flow machining (AFM) process and its limitations is required before it is presented in combination with a magnetic field assistive component.

4.4.2.1 Introduction and Limitations of AFM

Abrasive flow machining is an advanced micro-/nano-finishing technique extensively used in aerospace, the automotive industry, tool and die making, electronic parts and medical component manufacturing. AFM is used to deburr, finish and polish micro- to macro-size-machined components; even those having inaccessible areas, complex internal passages in 3D complex parts such as airfoils, turbine blades, fuel spray nozzles, discs, shafts, dies, impeller blades and gears [30]. This ultra-finishing technique is capable to produce nano-metre scale surface finish in a wide range of materials including stainless steels, metal matrix composites (MMCs) and superalloys.

Finishing occurs by extruding a deformable plug (semi-viscous state) made up of a mixture of abrasives and carrier medium through the cavities to be machined. The plug effectively removes protrusions along the cavity walls while removing the most material where the fluid flow is restricted. The plug acts as a tool with random cutting edges with an infinite orientation of abrasive media effectively removing workpiece material in the form of microchips [30, 31].

The normal forces (pressure induced) exerted by the abrasive particles on the workpiece are mainly responsible for the finishing rate and surface finish obtained. It is not possible to control these forces in conventional AFM. In order to overcome this limitation and to improve the performance in terms of enhanced material removal rate and surface quality, it is hybridized by the addition of a magnetic field. In this hybrid AFM process, the abrasive force acting on the workpiece surface is controlled externally by changing the magnetic flux density. This may be accomplished, depending on the set-up, by varying the electric current in an electromagnet coil or by changing the working gap when a permanent magnet is used. Magnetic field-assisted abrasive flow machining (MAAFM) is an important hybrid AFM process where the assistance of a magnetic field provides the means of controlling the cutting forces and consequently the outcome of the process includes higher surface finish and finishing rate. This process is capable of producing a surface roughness of 8 nm or lower [32].

4.4.2.2 Details of MAAFM

Magnetic field-assisted abrasive flow machining also known as magnetic abrasive finishing is a hybrid AFM process developed by applying a magnetic field around the workpiece to thereby control the magnitude of the cutting forces. This provides the means to generate the required micro-/nano-finish on external and internal surfaces of flat and cylindrical shapes made of magnetic and non-magnetic materials.

In this hybrid AFM process, ferromagnetic or magnetic abrasive particles (sintered mixture of ferromagnetic particles, i.e. iron particles, and fine abrasive particles, i.e. Al_2O_3, SiC, CBN or diamond) are used for finishing action under the pressure applied by the magnetic field.

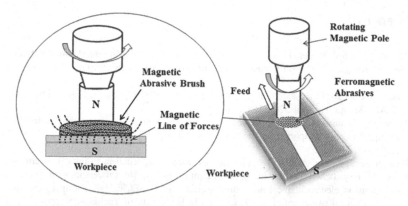

Fig. 4.12 Mechanism of finishing in magnetic field-assisted abrasive flow machining [32], Reproduced with kind permission from Elsevier

The basic operating principle of MAAFM is presented in Fig. 4.12. Finishing occurs by applying the ferromagnetic abrasive particles under the influence of a magnetic field in the machining gap between the workpiece top surface and bottom face of the rotating electromagnet. As the electromagnet is energized, the ferromagnetic particles are aligned along the magnetic lines of force. This effectively means that the particles are coalesced in lines between the poles to act as flexible magnetic abrasive brush. This magnetic brush has multiple cutting edges which help to remove material from the workpiece in the form of minute chips [32, 33]. The material removal from the workpiece surface occurs by the action of two forces, i.e. the normal component of the magnetic force is responsible for the abrasive particles to penetrate the workpiece surface while the rotating magnetic field induces rotation of the magnetic abrasive brush that removes workpiece material in the form of microchips. Since the magnitude of the machining force (induced by the magnetic field) is small, material removal rates are low and a mirror-like surface finish (i.e. average roughness values in the nano-metre range) can be obtained [33].

The important process parameters affecting the performance of MAAFM are size and type of the magnetic abrasives, mixing ratio of abrasive grains with magnetic particles, rotational speed of the electromagnet, magnetic flux density, working gap and workpiece material properties. Moreover by intelligent control of the excitation of the electromagnet (using a continuous or pulsed DC power supply) and thereby the electromagnetic behaviour of the magnetic abrasive particles may increase the material removal rate and improve the surface quality.

Magnetic abrasive deburring, ultrasonic-assisted magnetic abrasive finishing, electrolytic magnetic abrasive finishing, and pulsating current magnetic abrasive finishing are some important variants of MAAFM. This process has been successfully used to finish external surfaces and internal surfaces of cylindrical workpieces [34–36] and for micro-deburring of drilled holes using a permanent magnet in place of the conventionally used electromagnet [37].

References

1. Colwell L (1956) The effects of high-frequency vibrations in grinding. Trans ASME 78:837
2. Isaev A, Anokhin V (1961) Ultrasonic vibration of a metal cutting tool. Vest Mashinos (in Russian)
3. Skoczypiec S (2011) Research on ultrasonically assisted electrochemical machining process. Int J Adv Manuf Technol 52(5):565–574
4. Ruszaj A, Zybura M, úrek R, Skrabalak G (2003) Some aspects of the electrochemical machining process supported by electrode ultrasonic vibration optimization. J Eng Manuf 217:1365–1371
5. Skoczypiec S, Ruszaj A (2007) Application of ultrasonic vibration to improve technological factors in electrochemical machining of titanium alloys. In: Proceedings of international symposium on electrochemical machining technology INSECT 2007 (Scripts precision and micro production engineering), vol 1. Cheminitz University of Technology, pp 143–148
6. Pa PS (2007) Electrode form design of large holes of die material in ultrasonic electrochemical finishing. J Mater Process Technol 470–477
7. Ghoshal B, Bhattacharyya B (2013) Influence of vibration on micro-tool fabrication by electrochemical machining. Int J Mach Tool Manuf 64:49–59
8. Endo T, Tsujimoto T, Mitsui K (2008) Study of vibration-assisted micro-EDM–the effect of vibration on machining time and stability of discharge. Precis Eng 32:269–277
9. Chavoshi SZ, Luo X (2015) Hybrid micro-machining processes: a review. Precis Eng 41:1–23
10. Kumar MN, Subbu SK, Krishna PV, Venugopal A (2014) Vibration assisted conventional and advanced machining: a review. Procedia Eng 97:1577–1586
11. Abdullah A, Shabgard M, Ivanov A, Shervanyi M (2008) Effect of ultrasonic- assisted EDM on the surface integrity of cemented tungsten carbide (WCCo). Int J Adv Manuf Technol 41:268–280
12. Abdullah A, Shabgard M (2008) Effect of ultrasonic vibration of tool on electrical discharge machining of cemented tungsten carbide (WC-Co). Int J Adv Manuf Technol 38:1137–1147
13. Wansheng Z, Zhenlong W, Shichun D, Guanxin C, Hongyu W (2002) Ultrasonic and electric discharge machining to deep and small hole on titanium alloy. J Mater Process Technol 120 (1–3):101–106
14. Uhlmann E, Domingos DC (2013) Investigations on vibration-assisted EDM-machining of seal slots in high-temperature resistant materials for turbine components. Procedia CIRP 6:71–76
15. Rajurkar KP, Kozak J (2001) Laser assisted electrochemical machining. Trans NAMRI 29:421–427
16. Pajak PT, Desilva AKM, Harrison DK, Mcgeough JA (2006) Precision and efficiency of laser assisted jet electrochemical machining. Precis Eng 30:288–298
17. Desilva AKM, Pajak PT, Mcgeough JA, Harrison DK (2011) Thermal effects in laser assisted jet electrochemical machining. Ann CIRP 60:243–246
18. Zhang H, Xu J (2010) Modelling and experimental investigation of laser drilling with jet electrochemical machining. Chin J Aero 23:454–460
19. Kuo CL, Huang JD, Liang HY (2003) Fabrication of 3D metal microstructures using a hybrid process of micro-EDM and laser assembly. Int J Adv Manuf Technol 21:796–800
20. Kim S, Kim BH, Chung DK, Shin HS, Chu CN (2010) Hybrid micromachining using a nano second pulsed laser and micro EDM. J Micromech Microeng 20:015037–015038
21. Li L, Diver C, Atkinson J, Wagner RG, Helml HJ (2006) Sequential laser and EDM micro-drilling for next generation fuel injection nozzle manufacture. Ann CIRP 55:179–182
22. Takhata K, Aoki S, Sato T (1996) Fine surface finishing method for 3-D microstructures. In: Proceedings of MEMS 96, pp 73–78
23. Zhu D, Zeng YB, Xu ZY, Zhang XY (2011) Precision machining of small holes by the hybrid process of electrochemical removal and grinding. Ann CIRP 60:247–250

24. Kozak J, Skrabalak G (2014) Analysis of abrasive electrochemical grinding process (AECG). In: Proceedings of the world congress on engineering 2014 Vol II, WCE 2014, London, UK, 2–4 July 2014
25. Lin YC, Chen YF, Wang AC, Sei WL (2012) Machining performance on hybrid process of abrasive jet machining and electrical discharge machining. Trans Met Soc China 22:775–780
26. Shufeng S, Shiming JI, Depeng T, Wei X, Xin W (2012) Abrasive assisted EDM and ECM compound machining. J Mech Eng 17:159–164
27. Lin YC, Lee HS (2008) Machining characteristics of magnetic force-assisted EDM. Int J Mach Tool Manuf 48:1179–1186
28. Jahan MP (2013) Micro-electrical discharge machining. In: Davim JP (ed) Nontraditional machining processes: research advances. Springer, London, pp 111–152
29. Yeo SH, Murali M, Cheah HT (2004) Magnetic field assisted micro electro-discharge machining. J Micromech Microeng 14:1526–1529
30. Jain VK (2013) Micromanufacturing processes. Taylor and Francis LLC, Florida
31. Rhoades LJ (1988) Abrasive flow machining. Manuf Eng 75–78
32. Jain VK (2009) Magnetic field assisted abrasive based micro-/nano-finishing. J Mater Process Technol 209:6022–6038
33. Singh DK, Jain VK, Raghuram V (2005) On the performance analysis of flexible magnetic abrasive brush. Mach Sci Technol 9:601–619
34. Jain VK, Kumar P, Behera PK, Jayswal SC (2001) Effect of working gap and circumferential speed on the performance of abrasive finishing process. Wear 250:384–390
35. Komanduri R (1996) On material removal mechanisms in finishing of advanced ceramics and glasses. Ann CIRP 45(1):509–514
36. Kim JD (1997) Development of a magnetic abrasive jet machining system for internal polishing of circular tubes. J Mater Process Technol 71:384–393
37. Madarkar R, Jain VK (2007) Investigations into magnetic abrasive microdeburring (MAMDe). In: Anil B, Radhakrishnan V, Sunilkumar K (eds) Proceedings of the conference COPEN. Allied Publishers, New Delhi, pp 307–312

Index

© The Author(s) 2016
K. Gupta et al., *Hybrid Machining Processes*,
Manufacturing and Surface Engineering,
DOI 10.1007/978-3-319-25922-2

Printed in the United States
By Bookmasters